SALMONELLA

A practical approach to the organism and its control in foods

Chris Bell
Consultant Food Microbiologist
UK

Alec Kyriakides
Company Microbiologist
Sainsbury's Supermarkets Ltd
London, UK

Blackwell
Science

© 2002 C. Bell and A. Kyriakides

Blackwell Science Ltd
Editorial Offices:
Osney Mead, Oxford OX2 0EL
25 John Street, London WC1N 2BS
23 Ainslie Place, Edinburgh EH3 6AJ
350 Main Street, Malden
  MA 02148-5018, USA
54 University Street, Carlton
  Victoria 3053, Australia
10, rue Casimir Delavigne
  75006 Paris, France

Other Editorial Offices:

Blackwell Wissenschafts-Verlag GmbH
Kurfürstendamm 57
10707 Berlin, Germany

Blackwell Science KK
MG Kodenmacho Building
7-10 Kodenmacho Nihombashi
Chuo-ku, Tokyo 104, Japan

Iowa State University Press
A Blackwell Science Company
2121 S. State Avenue
Ames, Iowa 50014-8300, USA

First published 2002

Set in 10.5/12.5 pt Garamond Book
by DP Photosetting, Aylesbury, Bucks
Printed and bound in Great Britain by
MPG Books Ltd, Bodmin, Cornwall

DISTRIBUTORS

Marston Book Services Ltd
PO Box 269
Abingdon
Oxon OX14 4YN
(*Orders:* Tel: 01235 465500
            Fax: 01235 465555)

USA and Canada
Iowa State University Press
A Blackwell Science Company
2121 S. State Avenue
Ames, Iowa 50014-8300
(*Orders:* Tel: 800-862-6657
            Fax: 515-292-3348
            Web: www.isupress.com
            email: orders@isupress.com)

Australia
Blackwell Science Pty Ltd
54 University Street
Carlton, Victoria 3053
(*Orders:* Tel: 03 9347 0300
            Fax: 03 9347 5001)

A catalogue record for this title is available
from the British Library

ISBN 0-632-05519-7

Library of Congress
Cataloging-in-Publication Data
Bell, C., 1946–
    Salmonella: a practical approach to the
    organism and its control in foods/Chris Bell,
    Alec Kyriakides.
         p. cm.
    Includes bibliographical references and
    index.
    ISBN 0-632-05519-7 (alk. paper)
    1. Salmonellosis. 2. Salmonella.
    3. Food—Microbiology. I. Kyriakides,
    Alec. II. Title.

QR201.S25 .B455 2001
616.9′27—dc21   00-066804

For further information on
Blackwell Science, visit our website:
www.blackwell-science.com

CONTENTS

Contents

FOREWORD

This book provides an extensive review of this important human pathogen and is not only highly informative but is also an enjoyable read. The authors wisely chose to illustrate the importance of *Salmonella* as a foodborne pathogen and its behaviour not only by reference to published work on scientific studies but also by discussing a range of food poisoning outbreaks and food contamination incidents. The book provides a wealth of information that is of relevance to a range of scientific disciplines and will be particularly interesting to those with an interest in food production and food safety and also in public health microbiology.

When one reads this book, it becomes clear that *Salmonella* spp. are really quite remarkable organisms that have the capability of responding to and surviving in a wide variety of different foods and environments. The book also gives a good indication of the ubiquity of this important human pathogen and also the very wide range of food types that it has contaminated in the past and that will surely become contaminated in the future. The control of salmonellosis requires vigilance by all those involved in the food chain, from farmer to consumer, and this valuable book provides information that will be of help to those involved in the production of safe food and also people intending to enter this field.

T.J. Humphrey
Professor of Food Safety
Department of Clinical Veterinary Science
University of Bristol

The Practical Food Microbiology Series has been devised to give practical and accurate information about specific organisms of concern to public health. The titles in the series are:

E. coli
Listeria
Clostridium botulinum
Salmonella

1

BACKGROUND

INTRODUCTION

The genus name *Salmonella* was first suggested by Lignières in 1900 in recognition of the work carried out by the American bacteriologist, D.E. Salmon, who, with T. Smith in 1886, described the hog cholera bacillus causing 'swine plague' that they named *Bacterium suipestifer* (Topley and Wilson, 1929a; D'Aoust, 1989). This organism became the type species of *Salmonella*, *S. choleraesuis*. However, even as late as 1929, Topley and Wilson (1929a) referring to the new genus *Salmonella* expressed doubt as to whether the genus would attain official recognition because 'the *Salmonella* of one author is not yet that of another'. They did acknowledge that the name 'salmonella' was useful as a designation for 'non-lactose-fermenting bacilli which form acid and gas in dextrose and other sugars, ferment xylose and give an alkaline reaction in litmus milk'. They defined food poisoning as 'a disease characterised by acute gastro-enteritis of sudden onset and of comparatively brief duration, which not infrequently can be traced to the consumption of food infected with an organism of the salmonella group – usually *Bacterium enteritidis* or *Bacterium aertrycke*' (Topley and Wilson, 1929b).

In 1888, Gaertner isolated *Bacterium enteritidis* (later re-named *S. enteritidis*) from both the meat of an emergency-slaughtered cow and the organs of a man who was one of 58 people who consumed the meat and developed food poisoning. The man, who died, had eaten about 1.5 lb of the meat and died in 36 hours. This is probably the first laboratory confirmed outbreak of salmonellosis (Topley and Wilson, 1929b). By the early 1900s, some other important species of *Salmonella* such as *typhosum* (later typhi), *paratyphosum* A and B (*paratyphi* A and B), *gallinarum* and *typhimurium* had been characterised and the role of members of the *Salmonella* group of organisms as agents of disease in man and animals was becoming widely recognised. Today, *Salmonella* species are recognised as very important foodborne and waterborne organisms and the cause of a significant range of illnesses including food poisoning (gastroenteritis), typhoid (enteric fever), paratyphoid, bacteraemia, septicaemia and a variety of sequelae.

To date, almost 2400 serotypes (commonly regarded as species) of *Salmonella* have been identified and they are of major concern to nearly all sectors of the food industry. Indeed, the organism is included as a pathogen of concern in many food industry manufacturing and buying specifications as well as in standards in food legislation.

This book aims to provide the reader with information about the nature of the hazard presented by *Salmonella* to a range of food products and the means for controlling these organisms.

TAXONOMY OF *SALMONELLA*

Up until the middle of the twentieth century, the generic terms *Bacterium* or *Bacillus* (both meaning stick) were used to describe the broad group of Gram-negative, non-sporing rods occurring in the intestinal tract of man and animals, on plants and in the soil, and leading either a saprophytic, commensal or pathogenic existence. *Salmonella* was included in this group. By the 1960s, the name *Salmonella* was widely accepted to delineate a specific genus of the family Enterobacteriaceae and it was included in the *Approved Lists of Bacterial Names* published in 1980. *Salmonella* are facultatively anaerobic, Gram-negative, straight, small (0.7–1.5×2.0–$5.0\,\mu m$) rods, which are usually motile with peritrichous flagella. Table 1.1 shows some key biochemical characteristics of *Salmonella* and Figure 1.1 shows the DNA relatedness of *Salmonella* to some other genera of the Enterobacteriaceae. *Salmonella* are antigenically distinguishable by agglutination (formation of aggregates/clumps) reactions with homologous antisera. Work on their antigenic variation commenced over 80 years ago with significant contributions being made during the 1920s. Such work was based on the differences in the antigens on the surface of the bacterial cell where present, i.e. O, the somatic or outer membrane antigens; H, the flagella antigens, which in many *Salmonella* cultures are expressed in alternate phases of two different antigenic types (H1 and H2); and in addition, a very few produce Vi, the capsular antigens. The combination of antigens, referred to as the antigenic formula, is unique to each *Salmonella* serotype (Old, 1992). By the end of the 1920s, approximately 20 serotypes had been identified, and the names given to each different type (species) generally related to the

Table 1.1 Biochemical characteristics of *Salmonella*, adapted from Brenner (1984) and Le Minor (1984)

Characteristic	Usual reaction
Catalase	+
Oxidase	−
Acid produced from lactose	−
Gas produced from glucose*	+
Indole	−
Urease produced	−
Hydrogen sulphide produced from triple-sugar iron agar	+
Citrate utilised as sole carbon source*	+
Methyl red	+
Voges–Proskauer	−
Lysine decarboxylase	+
Ornithine decarboxylase	+

+ = positive reaction; − = negative reaction.
*An important exception is Typhi which is negative in these tests.

disease caused to the host from which the organism was isolated or after the place in which it was first isolated. This latter approach became the common convention for naming new types of *Salmonella*. Most of the currently recognised serotypes, almost 2400 of them and increasing in number every year, cannot be distinguished by biochemical tests, and separation is based purely on differences in the antigenic formulae.

The identification and labelling of antigens initiated by P.B. White in 1926 was further developed by F. Kauffmann in the early 1930s. The now famous (in microbiological circles) Kauffmann–White Scheme detailing the antigenic formulae and grouping of each *Salmonella* serotype was adopted for general use in 1934 by a special sub-committee of the International Association of Microbiologists (Wilson and Miles, 1964) and has been in use ever since. As the number of serotypes proliferated and each was accorded equal 'species' status despite their doubtful clinical significance, the fact that most of the serotypes were not able to be differentiated by biotyping led to concerns and doubts about the appropriateness of the nomenclature. As a consequence, some alternatives were suggested and actually used by workers in different countries (Old, 1992) and one particular style of nomenclature proposed by Ewing (1972) was used in the USA for some years.

Phage typing is a valuable addition to the methods for differentiating strains of *Salmonella* serotypes and was first used to differentiate strains

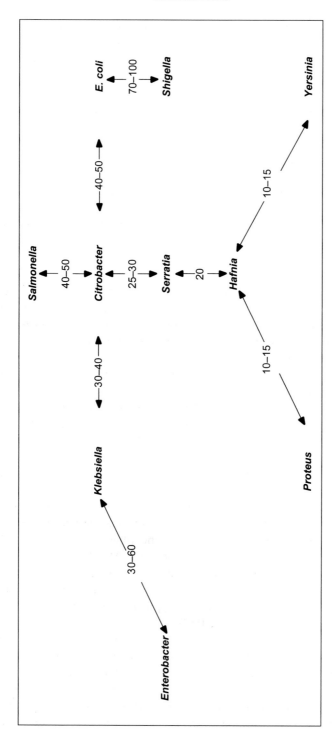

Figure 1.1 DNA relatedness of *Salmonella* and some other genera of the Enterobacteriaceae (numbers represent approximate percentage relatedness), adapted from Brenner (1984).

of *S. typhi*. Phage type is determined from the sensitivity of cells to the lytic activity of selected bacteriophages. These methods are now applied to other serotypes of *Salmonella*, particularly those of importance to human health including *S. paratyphi* B, *S. typhimurium*, *S. enteritidis*, *S. agona*, *S. hadar* and *S. virchow*. Phage type nomenclature includes reference to provisional phage type (PT) and definitive phage type (DT) indicating the status of the designation. Because of the fast growing number of types being described between the 1950s and 1970s, it was considered premature to finalise designations, so provisional numbers were allocated (PT). From the 1970s onwards, with the greater under-standing and experience gained through the various schemes in place, definitive numerical designations (DT) were able to be used (Anderson *et al.*, 1977; D'Aoust, 1989; Old and Threlfall, 1998). Phage typing is now a key characteristic used in epidemiological studies and in tracing food-borne outbreaks of salmonellosis.

Over the last 20 years, the results from DNA relatedness studies have indicated that all *Salmonella* serotypes are probably a single bacterial species and now, by international agreement, this is called *Salmonella enterica*. Seven subspecies have been identified within the species and one of these, *S. enterica* subsp. *bongori*, has been found to be suffi-ciently different from the other subgroups to be regarded as a separate species proposed to be called *S. bongori* (Threlfall *et al.*, 1999). Table 1.2 indicates the current nomenclature applied to *Salmonella* and some typical biochemical reactions that can help differentiate the subspecies. The subspecies of most concern in relation to food safety is *S. enterica* subsp. *enterica*; >99% of *Salmonella* isolated from man belong to this subspecies (Old, 1992). *Salmonella enterica* subsp. *arizonae* is also occasionally of concern to food safety. Under this new system of nomenclature, serotype names are only used now to distinguish between members of *S. enterica* subsp. *enterica* (Table 1.2) and serotype names are accorded similar status as biotyping and phage typing in the differ-entiation of *Salmonella* within a subspecies. The formal names now accorded to *Salmonella* are somewhat cumbersome, e.g. *S. enterica* subsp. *enterica* serotype Typhimurium (formerly *S. typhimurium*), *S. enterica* subsp. *enterica* serotype Paratyphi B biotype Java (formerly *S. java*) (Old, 1992; Threlfall *et al.*, 1999). For practical purposes, these are shortened to *S.* Typhimurium or *S.* Paratyphi B biotype Java, respectively. For the purposes of this book, although drawing on many reports that have used the Kauffmann–White Scheme of *Salmonella* nomenclature, this current practical nomenclature style of serotype names will be used together with any additional descriptive information that is supplied for some of the more predominant serotypes, e.g. provisional phage type

Table 1.2 Classification of *Salmonella* and some differentiating characteristics between subspecies, adapted from Le Minor and Popoff (1987), Rowe and Hall (1989) and Old and Threlfall (1998)

Salmonella nomenclature	*S. enterica* subsp. *enterica*	*S. enterica* subsp. *salamae*	*S. enterica* subsp. *arizonae*	*S. enterica* subsp. *diarizonae*	*S. enterica* subsp. *houtenae*	*S. enterica* subsp. *indica*	*S. bongori*
DNA hybridisation subgroup	I	II	IIIa Monophasic serovars	IIIb Diphasic serovars	IV	VI	V
Corresponding Kauffmann subgenera	I	II	III	III	IV		V
Designation	Majority of serotypes are in this subspecies Serotypes are named	Designated by their antigenic structure only – not named	Formerly known as *Arizona*		Designated by their antigenic structure only – not named	Serotypes previously included in subgroup/ subgenus I or II	Formerly regarded as atypical serotypes of subgroup/ subgenus I
Occurrence	Occur in humans and animals	Common in reptiles, rare in humans			Rare; mainly found in reptiles		

Table 1.2 *Continued*

Salmonella nomenclature	S. enterica subsp. enterica	S. enterica subsp. salamae	S. enterica subsp. arizonae	S. enterica subsp. diarizonae	S. enterica subsp. boutenae	S. enterica subsp. indica	S. bongori
Examples	Agona Dublin Enteritidis Poona Typhi Typhimurium Virchow	Gilbert Helsinki Makoma Neasden Sofia Tranaroa	Arizonae	Arizonae	Argentina Houten Marina Roterberg Wassenaar	Bornheim Ferlac Srinagar	Balboa Bongor Brookfield Malawi
Biochemical reactions:							
Dulcitol	+	+	–	–	–	d	+
Lactose	–	–	–	+	–	d	–
O.N.P.G.	–	–	+	+	–	d	+
Salicin	–	–	–	–	+	–	–
Sorbitol	+	+	+	+	+	–	+
Malonate	–	+	+	+	–	–	–
d-tartrate	+	–	–	–	–	–	–
Mucate	+	+	+	d	–	+	+
Gelatinase	–	+	+	+	+	+	–
β-glucuronidase	d	d	–	+	–	d	–
Galacturonate	–	+	–	+	+	+	+
Growth in KCN	–	–	–	–	+	–	+

+ = > 90% strains positive.
– = > 90% strains negative.
d = different reactions.

(PT) or definitive phage type (DT). Therefore, the new nomenclature will be used from this point.

ILLNESS CAUSED BY *SALMONELLA*

From the time of probably the first laboratory confirmed outbreak of salmonellosis in 1888, *Salmonella* have been considered to be the most important of causal agents of foodborne illness throughout the world and hundreds of outbreaks of foodborne salmonellosis still occur in most countries every year.

Incidents can be sporadic, affecting individuals, but outbreaks are common and, on occasion, involve very large numbers of cases, e.g. over 200 000 cases in a single outbreak attributable to a single source (Hennessy *et al.*, 1996). In developed countries where food-poisoning statistics have been collected and collated for many years, salmonellosis is recognised as the most important foodborne illness occurring in outbreaks. More cases of *Campylobacter* enteritis are generally recorded per year than for salmonellosis but these are more often sporadic cases rather than large community outbreaks.

Table 1.3 shows the numbers of *Salmonella* infections recorded by the Public Health Laboratory Service for England and Wales between 1986

Table 1.3 Incidence of reported cases of salmonellosis for England and Wales 1986–1999, adapted from Anon (1999a and 2000a)

Year	Reported numbers of cases	Reference
1986	16 976	Anon (1999a)
1987	20 532	
1988	27 478	
1989	29 998	
1990	30 112	
1991	27 693	
1992	31 355	
1993	30 650	
1994	30 411	
1995	29 314	
1996	28 983	
1997	32 596	
1998	23 728	Anon (2000a)
1999	17 532	
2000	14 845 provisional	

and 2000. The highest level recorded was in 1997 when the number of reports exceeded 32 000 but it has been well over 20 000 per year since 1988. In recent years, these figures have been decreasing as a result of public health and industry measures to control the organism, particularly in raw foods such as eggs and poultry. Reported cases have always been regarded as only a guide to the real number of cases occurring, variably discussed as between 10 times and 100 times the reported numbers. A recently completed survey of infectious intestinal disease conducted in England and Wales demonstrated that for every confirmed case of salmonellosis approximately 3.2 cases occurred in the community (Wheeler *et al.*, 1999; Anon, 2000b). Whatever the real figure, salmonellosis is clearly an important public health issue and the role of food in its dissemination must be regarded by the food industry as evidence of the need to give priority to control of the causative organism.

Commercially produced foods, food produced by private caterers for special events, and domestically produced foods including those in hotels, homes for the elderly, hospitals, schools and individual households have all been linked to outbreaks, large and small, of salmonellosis. Most, if not all, of these should have been preventable. In this book we focus on commercially produced foods and the control of *Salmonella* in commercial production systems.

Some *Salmonella* serotypes are specifically host-related, e.g. Typhi and Paratyphi A are associated with human hosts and Gallinarum with poultry, whereas some are more particularly infectious to some animals, e.g. Choleraesuis to pigs and Dublin to cattle, but may still cause infections in humans. The term host-restricted has been advocated for *Salmonella* serotypes that are almost exclusively associated with a particular host species, e.g. *S.* Gallinarum and *S.* Typhi (Uzzau *et al.*, 2000). Serotypes that are commonly associated with one host species but can cause illness in other species are referred to as host-adapted, e.g. *S.* Dublin and those that cause illness in unrelated species are given the term unrestricted.

Salmonella species cause illness by means of infection, i.e. the organism grows and multiplies in the host's body and becomes established in or on the cells or tissue of the host. *Salmonella* multiply in the small intestine, colonising and subsequently invading the intestinal tissues, producing an enterotoxin and causing an inflammatory reaction and diarrhoea (see D'Aoust, 1997 for detailed information on the pathogenicity of *Salmonella*). When *Salmonella* overcomes the host's natural defence systems, the organisms can get into the blood stream and/or the lymphatic system and cause more severe illnesses.

Illnesses caused by *Salmonella* serotypes range from gastroenteritis to enteric (typhoid) fever and septicaemia and chronic sequelae (Table 1.4). Young children, the elderly and those with underlying chronic illness or immuno-compromised individuals are particularly vulnerable to salmonellosis.

Enteric fever is a severe and debilitating illness and treatment relies on supportive therapy and antibiotic treatment to eliminate the infection. However, the development of antibiotic resistance by typhoid organisms to the antibacterials commonly used in the treatment of enteric fever (chloramphenicol, ampicillin and trimethoprim) is of growing concern in relation to the efficacy of treatment regimes particularly in developing countries where enteric fever is more common (Old and Threlfall, 1998).

In developed countries, normally healthy individuals recover from salmonellosis (mild to moderate gastroenteritis) with supportive treatment including fluid and electrolyte replacement and without recourse to antibiotics. Deaths resulting from salmonellosis in outbreaks are rare, but do occur. Use of antibiotics in these cases can increase the time during which the organism is excreted and so is not a recommended approach to treatment. In developing countries however, salmonellosis is more commonly a severe gastroenteritis with up to 40% of cases accompanied by septicaemia in some outbreaks and up to 30% mortality, and it is the multiple-antimicrobial agent-resistant (resistant to up to nine antimicrobial agents, including widely used compounds such as ampicillin, chloramphenicol, gentamicin, kanamycin, streptomycin, sulphonamides, tetracycline, trimethoprim, furazolidone and nalidixic acid) strains of *Salmonella* that are often involved in these outbreaks (Old and Threlfall, 1998). In particular, strains of *Salmonella* Typhimurium DT104, which are increasingly causing food-associated outbreaks of severe gastrointestinal infection (Besser *et al.*, 2000; Anon, 2000c), are commonly resistant to at least five and sometimes more antibiotics. Clearly, the increasing role of antibiotic-resistant *Salmonella* in cases of human illness is of considerable concern and is the subject of continuing study and discussion between scientists and governments (Anon, 1998a).

During recovery from enteric fever or salmonellosis, causative organisms may frequently persist in faeces during the convalescent period but gradually become less until organism excretion ceases. However, in a few cases, excretion may persist and the individual can become a chronic carrier of the organism, which may be excreted only intermittently. Such

Table 1.4 Illnesses caused by *Salmonella*, adapted from Anon (1995a) and International Commission on Microbiological Specifications for Foods (1996)

Illness	Infective dose	Characteristics of the illness	Serotypes involved
Gastroenteritis	Usually high numbers (>10 000 cells) required to cause illness but where organisms are protected, e.g. in high fat foods, low numbers (<100 cells) may cause illness	Incubation time 12–72h, commonly 12–36h. Lasts 2–7 days. Symptoms: diarrhoea (dehydration if this is severe), abdominal pain, vomiting, fever, sometimes fatal. Prolonged excretion may occur.	Mainly members of subgroup/ subgenus I but also, subgroup III
Enteric fever	Infective dose may be <1000 cells	Incubation time 7–28 days, average 14 days. Typhoid. High fever, malaise, nausea, abdominal pain, anorexia, delerium, constipation in the early stages, diarrhoea in the late stages. Convalescence may take up to 8 weeks. Carrier state can last several months to years.	*S.* Typhi and *S.* Paratyphi (subgroup/subgenus I)

(Continued on p. 12.)

Table 1.4 *Continued*

Illness	Infective dose	Characteristics of the illness	Serotypes involved
Septicaemia or bacteraemia		Caused when *Salmonella* are present in the blood stream. High fever, malaise, pain in the thorax and abdomen, chills and anorexia.	Members of subgroup/subgenus I
		Focal infections of some tissues may occur.	Members of subgroup/subgenus I particularly *S.* Typhimurium, *S.* Dublin and *S.* Choleraesuis
Sequelae		Uncommon. A variety have been identified including arthritis, osteoarthritis, appendicitis, endocarditis, pericarditis, meningitis, peritonitis and urinary tract infections.	Members of subgroup/subgenus I

individuals can be a significant hazard to food safety and public health. The classic case illustrating this issue is that of a woman nicknamed 'Typhoid Mary'; she was a cook in America who, in 1901, developed enteric fever and subsequently, via employment in different houses, was discovered to have been responsible for 26 cases of enteric fever over five years before being placed under observation by the New York Health Department. In 1914, she 'escaped' observation and was believed to have caused other outbreaks including 25 cases in a hospital, the cook being found to be 'Typhoid Mary', and one outbreak involving 1300 people. This case remains a salutary lesson concerning the potential dangers posed by long-term carriers of *Salmonella* and the need to follow public health advice concerning control of salmonellosis and typhoid/paratyphoid infections in food handlers (Anon, 1995a).

The serotypes of *Salmonella* responsible for most cases of reported salmonellosis do change over the years. Until the mid-1980s, *S.* Typhimurium was the most commonly reported serotype in the USA and UK, but *S.* Enteritidis has overtaken *S.* Typhimurium in more recent years due to a large rise, particularly in egg-related outbreaks and incidents of salmonellosis involving *S.* Enteritidis PT4. From time to time, reported cases involving different serotypes and/or phage types show an increase then decline, and careful epidemiological study is able to indicate sources, e.g. *S.* Enteritidis PT6 became the second most commonly reported phage type of *S.* Enteritidis in the UK in 1997 being responsible for over 1600 cases. This was attributed to shell eggs and poultry meat. Cases caused by multiple antibiotic-resistant *S.* Typhimurium DT104 have also increased in recent years and illness is reported to be more serious than that caused by other non-typhoid *Salmonella*. Meat products, raw milk and products made from raw milk are the main foods implicated in outbreaks attributed to *S.* Typhimurium DT104 but direct contact with livestock is also reported to be a risk factor for human infection.

It should be expected that yet other serotypes or phage types will 'emerge' as important causes of outbreaks for a time before they decline and others 'emerge'. Perhaps, the increasing spread of antibiotic resistance in the environmental 'pool' of *Salmonella* may lead to longer persistence of 'emerged' serotypes and an increase in numbers of types emerging.

The infective dose for causing foodborne salmonellosis in humans was, for decades, believed to be high, i.e. 100 000 to one million cells, but a number of outbreaks have now occurred in which the infective dose was found to be low, e.g. < 10–100 cells. This is especially the case where products containing a high fat level are involved, e.g. chocolate, cheese and salami (Table 1.5), and where the consumers are children or elderly people. Realising that such a low number of cells of *Salmonella* can be capable of causing illness serves to underline the importance of all the measures required to be taken to minimise the incidence and level of the organism at all stages of the food chain. Also, in order to provide reliable information, the microbiological test methods used to support HACCP (hazard analysis, critical control point) based food production safety and quality systems must be carefully selected and operated to ensure that they are suitable for detecting potentially low numbers in the sample types under examination.

The issues relating to the increasingly widespread antibiotic resistance of *Salmonella* and low infective dose, coupled with the ready availability of food raw materials from anywhere in the world and the use of more 'exotic' ingredients in innovative food product development provide a growing challenge to food microbiologists and technologists to devise and maintain controls of the persistent hazard of *Salmonella* in the food supply chain.

SOURCES OF *SALMONELLA*

The primary sources of *Salmonella* are the gastrointestinal tract of humans, domestic and wild animals, birds and rodents. Consequently, they are widespread in the natural environment including soil and waters in which they do not usually multiply significantly but may survive for long periods, i.e. many months in soil and dried animal faeces. Transmission routes to humans are mainly between humans, from animals via the food supply and from water and the environment. Foods contaminated by *Salmonella* usually look and smell normal; it is therefore essential that every effort is made by all those in the food production and supply industries to exclude the organism from foods wherever possible.

On occasion, multiple serotypes of *Salmonella* have been found associated with food raw materials implicated in outbreaks, e.g. 11 strains were identified in paprika powder including *S.* Saintpaul, *S.* Javiana, *S.* Rubislaw, *S.* Florian, *S.* Loenga and six un-named strains (Lehmacher *et al.*, 1995). Mung bean seeds, cocoa beans and coconut have also been variously reported to contain more than one serotype of *Salmonella*. O'Mahony *et al.* (1990) reported five serotypes isolated from

Table 1.5 Infective dose of *Salmonella* calculated to cause illness in selected outbreaks

Product	*Salmonella* serotype	Infective dose (no. of cells)	Reference
Cheddar cheese	Heidelberg	100–500	Fontaine *et al.* (1980)
Cheddar cheese	Typhimurium PT10	1 (<1–10)	D'Aoust (1989) and D'Aoust *et al.* (1985)
Chocolate balls	Eastbourne	<100	D'Aoust *et al.* (1975)
Chocolate bar	Napoli	<50	Greenwood and Hooper (1983)

bean sprouts. Investigation of a single outbreak affecting approximately 30 people between mid-November 1998 and March 1999 due to *S.* Paratyphi B variant Java (previously known as *S. java*) phage type Dundee (*c.* 26 cases) and phage type Battersea (4 cases), identified desiccated coconut as the source of the organisms. This was found to be contaminated with the main outbreak strain (Java phage type Dundee); a large number of the samples examined yielded this organism. Some of the infections caused by Java phage type Dundee were very serious resulting in bacteraemia and hospitalisation for some patients. A further 10 different serotypes of *Salmonella* were isolated from the contaminated coconut (Anon, 1999b and c; personal communication). As a result of this outbreak investigation, considerable quantities of desiccated coconut were withdrawn from sale and recalled from the public. Undoubtedly, a major and very serious outbreak was averted in this case.

An increasing number of outbreaks of salmonellosis in small children seem to be occurring due to poor supervision of hygiene during and following the handling of animals on farms and in schools, e.g. 32 cases of children between the ages of 2 and 4 years were diagnosed with salmonella enteritis possibly due to *S.* Enteritidis following exposure to chicks and ducklings in a classroom environment (Anon, 2000c). This is a salutary lesson to the nursery staff and parents and a sad lesson for the children, as well as a clear reminder to all that personal hygiene in the form of handwashing following any 'dirty' activity, including handling farm animals or pets, is essential prior to handling food or putting fingers or food in the mouth!

Enteric fever, as a disease, has been known for centuries and even as early as the late 1850s the roles of infected faeces, poor personal hygiene and faecally contaminated water and milk in the spread of typhoid were recognised (Wilson and Miles, 1964). However, the causal agent, the typhoid bacillus, was only first described in 1880 and isolated in 1884. By 1900, *S.* Typhi, *S.* Paratyphi A and *S.* Paratyphi B had all been recognised as causal agents of enteric fever (typhoid) (Fairbrother, 1938); *S.* Sendai being recognised later. *Salmonella* Typhi is responsible for the most severe form of typhoid.

Typhoid fever is now rare in developed countries but a few notable food-associated outbreaks have occurred, e.g. in 1936 a human carrier contaminated milk which caused over 700 cases and about 70 deaths in Bournemouth, Poole and Christchurch (UK) (Fairbrother, 1938; Wilson,

1942); in 1964, 507 cases and three deaths occurred in Scotland due to imported canned corned beef (Walker, 1965); in Germany in 1974, 417 cases and five deaths were linked to contaminated potato salad (International Commission on Microbiological Specifications for Foods, 1998) and in the USA in 1981, 80 cases were attributed to a contaminated meat product (Taylor *et al.*, 1984). The ultimate source of the causative organism in each case was human faecal contamination. In developing countries where sanitation facilities are poor and drinking water supplies are untreated, typhoid is relatively common. Poor hygiene and sanitation conditions facilitate person-to-person spread and water supplies contaminated with human faeces ensure that the typhoid causing organisms will remain as hazards from the environment and food contaminated from these sources.

The ubiquity of non-typhoid *Salmonella* makes them a persistent contamination hazard to all raw foods whether they are derived from animals, fish and shellfish, eggs, poultry and game birds, fruit, vegetables and salad vegetables, dairy produce, or cereals.

A number of surveys have been carried out over the years to monitor the incidence of *Salmonella* in foods and animal feedingstuffs (Tables 1.6 and 1.7). *Salmonella* serotypes have been reported in a variety of fresh produce products. A survey of fresh produce imported from 21 countries into the USA, carried out by the US Food and Drug Administration in 1999, found 3.5% (35/1003) samples positive for *Salmonella* and 1% (9/1003) positive for *Shigella* spp. (Food and Drug Administration, 2001a). *Salmonella* were detected on cantaloupe melons (8 samples), cilantro (16 samples), culantro (6 samples) and in one sample each of celery, lettuce, parsley, scallions and strawberries. Poultry and egg products have long been recognised as a source of *Salmonella*, e.g. 18 out of 23 flocks of poultry investigated in Denmark were found to be positive for *Salmonella* and seven different serotypes were isolated during the investigation (Skov *et al.*, 1999). Other studies have shown that multi-antibiotic resistant *S.* Typhimurium DT104 can also be found in poultry and the poultry growing environment (Rajashekara *et al.*, 2000). Although poultry products are not, at present, regarded as a major source of this particular phage type in relation to human infections, the potential for them to become so, is great. *Salmonella* have also been reported in the environment and products from aquaculture, *S.* Weltevreden being reported as the most commonly isolated serotype in shrimps imported into England and Wales and Australia (Dalsgaard, 1998).

Table 1.6 Incidence of *Salmonella* reported in foods

Food	Country	Percentage incidence* (number of samples)	Reference
Retail chicken	Sweden	0	ACMSF (1996)
Retail chicken	Portugal	48	
Retail chicken	Denmark	51	
Turkey	Netherlands	88.5 (26)	
Domesticated duck	Netherlands	80.1 (26)	
Wild duck	Netherlands	9.7 (62)	
Raw chicken, frozen (1987)	UK	64 (101)	ACMSF (1996)
Raw chicken, frozen (1990)	UK	54 (143)	
Raw chicken, frozen (1994)	UK	41 (281)	
Raw chicken, chilled (1987)	UK	54 (103)	
Raw chicken, chilled (1990)	UK	41 (143)	
Raw chicken, chilled (1994)	UK	33 (281)	
Raw shell eggs (packs of 6 eggs)	Northern Ireland	0.43 (2090 packs)	Wilson *et al.* (1998)
Unpasteurised milk	UK	0.45 (1097)	de Louvois and Rampling (1998)
Raw milk (from bulk tank trucks)	USA	4.7 (678)	McManus and Lanier (1987)
Raw cows' drinking milk	UK	0.06 (1591)	Anon (1998b)
Whole vegetables	Spain	7.5 (345)	Garcia-Villanova Ruiz *et al.* (1987)
Vegetables	USA	8 (50)	Beuchat (1996)
Artichoke	Spain	12 (25)	
Bean sprouts	Thailand	8.7 (344)	
Beet leaves	Spain	7.7 (52)	
Cabbage	Spain	17.1 (41)	
Cauliflower	Netherlands	7.7 (13)	
Cauliflower	Spain	4.5 (23)	

Table 1.6 *Continued*

Food	Country	Percentage incidence* (number of samples)	Reference
Celery	Spain	7.7 (26)	Beuchat (1996)
Lettuce	Netherlands	7.1 (28)	
Lettuce	Spain	6.3 (80)	
Lettuce	Italy	68.3 (120)	Ercolani (1976)
Fennel	Italy	71.9 (89)	
Raw sausage	UK 1990–91	8 (988)	Anon (1993a)
Raw meat, not poultry	UK 1990–91	3 (830)	
Offal	UK 1990–91	4 (458)	
Raw shellfish	UK 1990–91	2 (566)	Anon (1993a)
Crab & crab products	USA 1990–98	3.7 (298)	Heinitz et al. (2000)
Raw crustaceans other than crab		8.3 (4440)	
Dried/salted seafood		3.2 (792)	
Smoked fish/seafood		2.9 (344)	
Fin fish/skin fish		11.8 (2114)	
Prepared ready-to-eat seafoods		2.6 (2734)	
Molluscs, ready-to-eat, out-of-shell	UK	0.17 (2394)	Little et al. (1997)

*Method of sampling and quantity of food tested varies between studies and may affect the significance of the results. Primary publications should be consulted for full details of sampling procedures and test methods used.

Table 1.7 Incidence of *Salmonella* reported in some animal feedingstuffs and ingredients of animal feedingstuffs in the UK, adapted from ACMSF (1996)

Feedingstuff	Percentage incidence (number of samples)
Processed animal protein (UK and imported) sampled at the feed compounders (1994)	4.1 (6137)
Poultry compound feed (1994)	2.7 (14 256)
Ruminant compound feed (1994)	3.4 (3235)
Pig compound feed (1994)	3.6 (6598)
Protein concentrate (1994)	3.7 (1724)
Oilseed meals and other ingredients of animal feedingstuffs sampled at feed mills (1994)	4.9 (12 460)

Survey findings coupled to the wide variety of foods implicated in outbreaks of human salmonellosis (Table 1.8) further underlines the potential hazard these organisms represent in almost any food commodity area. The outbreaks listed in Table 1.8 are but a small representation of those that have occurred implicating different foods. For some food types, a large number of outbreaks have been recorded around the world, e.g. salad vegetable sprouts have been implicated in a number of significant outbreaks of salmonellosis in relatively recent years (Table 1.9) involving a variety of serotypes; 13 different serotypes were identified in the outbreak examples given in Table 1.9.

The development of 'Salm-Net' in 1994, subsequently named 'Enternet', has facilitated international collaboration in the surveillance and investigation of human salmonellosis and international outbreaks of this illness. This collaboration has helped in the identification of sources of the organism in major international outbreaks, e.g. the identification of a savoury snack causing an outbreak (over 2000 cases) in Canada, England and Wales, Israel and the USA (Killalea *et al.*, 1996; Anon, 2001).

As the food chain becomes even more globally complicated and international collaboration develops further, it should be anticipated that more 'new' foods will be linked to cases of salmonellosis. It is therefore, prudent to expect that *Salmonella* will be present in or on any raw food material, and to handle and process the material accordingly.

Table 1.8 Examples of food-associated outbreaks of illness caused by *Salmonella*

Salmonella serotype	Year	Country	Suspected food vehicle	Cases	Reference
Meat products					
Typhimurium PT8	1953	Sweden	Raw meat	8845	Lundbeck *et al.* (1955)
Typhi PT34	1964	Scotland	Corned beef	507	Walker (1965)
Newport	1981	Australia	Salami	279	Taplin (1982)
Goldcoast	1984	France	Pâté	506 (est. 5000)	Bouvet *et al.* (1986)
Enteritidis PT4	1984	England & Wales	Aspic glaze	766	Anon (1986); Palmer and Rowe (1986)
Virchow	1984	England & Wales	Ham	274	Anon (1986)
Agona PT15	1996	UK	Cooked turkey meat	9	Synnott *et al.* (1998)
Typhimurium DT124	1987–88	UK	Salami sticks	101	Cowden *et al.* (1989)
Typhimurium DT12	1989	UK	Cooked meats	545+	Sockett *et al.* (1993)
Typhimurium	1995	Italy	Salami	83	Pontello *et al.* (1998)
Milk products					
Heidelberg	1976	USA	Cheddar cheese	339+	Fontaine *et al.* (1980); D'Aoust (1989)
Typhimurium PT204	1981	Scotland	Raw milk	654	Cohen *et al.* (1983)
Typhimurium PT10	1984	Canada	Cheddar cheese	1500+	D'Aoust *et al.* (1985)
Typhimurium	1985	USA	Pasteurised milk	16 284 (est. > 150 000)	Lecos (1986); Ryan *et al.* (1987)
Typhimurium	1985	Switzerland	Vacherin Mont d'Or cheese	> 40	Sadik *et al.* (1986)
Ealing	1985	UK	Infant dried milk	76	Rowe *et al.* (1987)
Javiana and Oranienburg	1989	USA	Mozzarella	164	Hedberg *et al.* (1992)
Dublin	1989	UK	Unpasteurised milk soft cheese	42	Maguire *et al.* (1992)

(Continued on p. 22.)

Table 1.8 *Continued*

Salmonella serotype	Year	Country	Suspected food vehicle	Cases	Reference
Paratyphi B biotype Java	1993	France	Goats' milk cheese, unpasteurised	273	Desenclos *et al.* (1996); Threlfall *et al.* (1999)
Berta	1994	Canada	Unpasteurised milk soft cheese product	35+	Ellis *et al.* (1998)
Dublin	1995	Switzerland & France	French cheese (Doubs region)	25 France NR Switzerland	Vaillant *et al.* (1996)
Typhimurium 'DT12 atypical'	1997	France	Raw milk soft cheese - Morbier	113	De Valk *et al.* (2000)
Typhimurium DT104	1998	UK	Pasteurised milk	86	Anon (1998c)
Egg associated					
Typhimurium DT49	1988	UK	Mayonnaise	76+	Mitchell *et al.* (1989)
Enteritidis PT4	1991	UK	Custard in bakery goods	17	Barnes and Edwards (1992)
Enteritidis	1994	USA	Ice cream	est. 224 000	Hennessy *et al.* (1996)
Enteritidis PT4	1995	UK	Marshmallows	24	Lewis *et al.* (1996)
Enteritidis PT4	1999	Ireland	Egg fried rice	110	Cronin (1999)
Confectionery products					
Eastbourne	1973-4	Canada & USA	Chocolate	95 Canada c. 120 USA	Craven *et al.* (1975); D'Aoust *et al.* (1975)
Napoli	1982	England & Wales	Chocolate bars	245	Gill *et al.* (1983)
Nima	1985-86	USA & Canada	Chocolate coins (imported from Belgium)	33 (29 Canada, 4 USA)	Hockin *et al.* (1989)
Typhimurium	1987	Norway & Finland	Chocolate products	361 (349 Norway, 12 Finland)	Kapperud *et al.* (1990)

Table 1.8 *Continued*

Salmonella serotype	Year	Country	Suspected food vehicle	Cases	Reference
Fresh produce					
Saintpaul	1988	UK	Mung bean sprouts	143	O'Mahony *et al.* (1990)
Virchow PT34	1988	UK	Mung bean sprouts	7	O'Mahony *et al.* (1990)
Goldcoast	1989	UK	Mustard cress	14+	Joce *et al.* (1990)
Chester	1990	USA	Cantaloupe	295	Tauxe *et al.* (1997)
Javiana	1990	USA	Fresh tomatoes	176	Tauxe *et al.* (1997)
Poona	1991	USA	Cantaloupe	>400	Tauxe *et al.* (1997)
Montevideo	1993	USA	Tomatoes	100	Tauxe *et al.* (1997)
Bovismorbificans	1994	Finland	Sprouts	210	Puohiniemi *et al.* (1997)
Stanley	1995	USA & Finland	Alfalfa sprouts	242	Mahon *et al.* (1997)
Newport	1995–6	USA & Canada	Alfalfa sprouts	133	Van Beneden *et al.* (1999)
Saphra	1997	USA	Cantaloupe	>20	Anon (2000e)
Poona	2000	USA	Cantaloupe	>19	Anon (2000e)
Enteritidis	2000	USA	Mung bean sprouts	45	Anon (2000f)
Others					
Not reported	1955	Denmark	Mayonnaise	c. 10 000	ICMSF* (1980)
Typhi	1974	Germany	Potato salad	417	ICMSF* (1998)
Newport	1974	USA	Potato salad	3400 est.	Horwitz *et al.* (1977)
Typhimurium	1974	USA	Apple cider	>200	Anon (1975)
Enteritidis PT4	1977	Sweden	Mustard dressing	2865	Hellström (1980)
Oranienburg	1981–2	Norway	Black pepper	126	Gustavsen and Breen (1984)
Champaign	1988	Japan	Roast cuttlefish	330	D'Aoust (1997)
Manchester	1989	UK	Savoury corn snack (yeast based flavouring)	47	Joseph *et al.* (1991)

(Continued on p. 24.)

Table 1.8 *Continued*

Salmonella serotype	Year	Country	Suspected food vehicle	Cases	Reference
Agona PT15	1994	England & Wales, USA, Israel, Canada	Savoury snack	2000+	Killalea *et al.* (1996); Anon (2001)
Senftenberg	1995	England	Infant food, cereal	5	Rushdy *et al.* (1998)
Hartford and Gaminara	1995	USA	Unpasteurised orange juice	62	Cook *et al.* (1998)
Mbandaka	1996	Australia	Peanut butter	54	Ng *et al.* (1996)
Agona	1998	USA	Toasted oats cereal	209	Anon (1998d)
Typhimurium PT135A	1999	Australia	Unpasteurised orange juice	>400	Anon (1999d)
Muenchen	1999	USA & Canada	Unpasteurised orange juice	>300	Boase *et al.* (1999)
Java PT Dundee	1998–99	UK	Coconut	18	Anon (1999b and c)

NR = not recorded.

*ICMSF = International Commission on Microbiological Specifications for Foods.

Table 1.9 Examples of outbreaks of salmonellosis implicating salad vegetable sprouts

Country	Year	Type of sprout	*Salmonella* serotype	Reference
UK	1988	Bean sprouts	Saintpaul	O'Mahony *et al.* (1990)
UK	1989	Mustard cress	Goldcoast	Joce *et al.* (1990)
Finland & Sweden	1994	Alfalfa	Bovismorbificans	Puohiniemi *et al.* (1997)
Finland & USA	1995	Alfalfa	Stanley	Mahon *et al.* (1997)
Canada & Denmark	1995	Alfalfa	Newport	Aabo and Baggesen (1997)
USA (6 states) & Canada	1995–96	Alfalfa	Newport	Van Beneden *et al.* (1999)
USA	1996	Alfalfa	Montevideo/ Meleagridis	National Advisory Committee on Microbiological Criteria for Foods (1999)
Canada	1997	Alfalfa	Meleagridis	
USA	1997	Alfalfa	Infantis/Anatum	
USA	1997–98	Clover/alfalfa	Senftenberg	
USA	1998	Alfalfa	Havana/Cuba	
USA	1999	Alfalfa	Mbandaka	

2

OUTBREAKS: CAUSES AND LESSONS
TO BE LEARNT

INTRODUCTION

Salmonella probably causes more outbreaks of foodborne illness in the developed world than any other bacterium. Outbreaks have been recorded implicating all food groups including meat, poultry and eggs, dairy, confectionery, fruits and vegetables (Table 1.8) and these demonstrate the serious hazard presented by this organism.

It is incumbent on the food industry to use these outbreaks to identify lessons to be learnt and procedures that can be applied to prevent further damaging outbreaks from occurring.

Analysis of historical outbreaks identifies raw foods of animal origin together with cross-contamination to ready-to-eat foods through inadequate processing or personal hygiene as being significant contributory factors to the cause of outbreaks. Foods that meet the following criteria are far more likely to be implicated in outbreaks of salmonellosis:

- contain raw ingredients of animal origin
- raw ingredients subject to direct or indirect contamination from animal sources, i.e. manure or contaminated irrigation water
- raw ingredients not subsequently subjected to a bacterial destruction process
- products subject to post-process contamination
- process conditions that allow the growth of *Salmonella,* if present.

Continued outbreaks of salmonellosis serve as a reminder of the hazard presented by this organism and the need for constant vigilance in the manufacture, retail and consumer handling of foods. This chapter reviews a number of outbreaks that have occurred over the years involving a variety of foods and identifies what could have caused them and, importantly, what could have been done to prevent them. Information from outbreaks of foodborne illness is often incomplete and the authors have therefore used their own judgement in determining the critical factors likely to be involved. In this way it is hoped that production or handling mistakes or weaknesses that have allowed previous incidents and outbreaks to occur can be used to improve the safety of existing and future products, thereby preventing more illness being caused by this organism.

SPROUTED SALAD VEGETABLES: ENGLAND, SWEDEN, USA
AND CANADA

In March 1988, a cluster of salmonellosis infections caused by one *Salmonella* serotype was reported in the Oxford region of England (O'Mahony *et al.*, 1990). An epidemiological and microbiological investigation established a total of 143 cases in England caused by the outbreak strain. Seventy-three of the 143 cases were men, 67 women and three cases were recorded where the sex was not reported. Ninety-seven people who suffered illness were interviewed (80 primary cases, 17 secondary cases) plus one asymptomatic case. Of these, 95 reported suffering diarrhoea, 32 with blood in their stools. Forty-one individuals experienced nausea, 33 vomiting and 80 reported feverishness. Symptoms lasted from two to 42 days (median 10 days) and eight patients were admitted to hospital; there were no deaths.

The implicated food was bean sprouts (Table 2.1) with 48 out of the 80 primary cases who were interviewed recollecting that they had eaten bean sprouts in the week before onset of illness; 42 individuals reported eating the beans prouts without any further cooking (O'Mahony *et al.*, 1990). The bean sprouts were traced to at least five different producers, one of which was exclusively associated with 15 cases and, in association with other producers, was implicated in a further 14 cases. Six cases were linked to four other producers. The remainder of the cases could not be traced to an individual producer.

The outbreak strain was *Salmonella* Saintpaul which was isolated from affected individuals and packs of bean sprouts on retail sale from one of the producers. This strain was also isolated from bean sprouts and environmental samples taken from one producer implicated in a large number of the cases. Interestingly, *Salmonella* Virchow PT34 was also isolated from this source and this strain was subsequently found to be associated with seven cases of infection. *Salmonella* Virchow PT34 was also isolated from bean sprouts remaining in the implicated pack retrieved from the home of one affected individual.

Salmonella Saintpaul was isolated from a retail pack of bean sprouts produced by a manufacturer who was not implicated in any cases associated with this outbreak. *Salmonella* Enteritidis and *S.* Mbandaka were also recovered from waste material retrieved from this producer's premises. Retail packs of bean sprouts sampled by Public Health Laboratories in the north of England revealed contamination with a variety of other *Salmonella* serotypes including *S.* Lancing, *S.* Litchfield and *S.* Arizona,

Table 2.1 Outbreak overview: sprouted salad vegetables

Product types:	Mung bean sprouts, mustard cress, alfalfa sprouts
Years:	1988, 1989, 1995
Countries:	England/Sweden, England, USA
Levels:	Not reported
Organisms:	*S.* Saintpaul, *S.* Goldcoast and *S.* Newport
Cases:	143, 14+, 133 (no deaths)

Possible reasons

(i)   Use of contaminated seeds
(ii) Survival of *Salmonella* in seed washing/sanitation process (up to 7500 ppm chlorine)
(iii) Growth of *Salmonella* during germination and sprouting
(iv) Consumption of product without further cooking

Control options*

(i)   Source seed from a reputable supplier operating good agricultural practices in seed production and who is included in a supplier quality assurance programme supported by extensive intake testing of the seed batches together with trial germination
(ii) Chlorinated wash of seed with high chlorine levels (up to 20 000 ppm) supported by *Salmonella* testing of the spent water during production or an alternative decontamination procedure, e.g. using heat
(iii) Chlorinated wash of salad vegetables after germination
(iv) Cooking of vegetable sprouts

* Suggested controls are for guidance only and may not be appropriate for individual circumstances. It is recommended that a proper hazard analysis be carried out for every process and product to identify where controls must be implemented to minimise the hazard from *Salmonella*.

although these were not associated with any recorded illnesses. Samples of previously unopened sacks of mung bean seeds retrieved from the producer associated with a large number of cases were found to be contaminated with *S.* Saintpaul and *S.* Muenchen; this seed was imported from Australia in 1987.

In April 1988 the Department of Health issued advice to the public to boil bean sprouts for 15 seconds prior to consumption and a product recall, i.e. withdrawal from the marketplace was effected of all bean sprouts from the producer associated with many of the cases.

It is interesting to note that in the spring of 1988, a large outbreak of salmonellosis was reported in Sweden associated with the consumption of bean sprouts (Table 2.1); the outbreak strain was *S.* Saintpaul of the same antigenic description as the English outbreak strain. *Salmonella* Havana and *S.* Muenchen were also reported to have caused infections in this outbreak (O'Mahony *et al.*, 1990) and were isolated from the bean sprouts implicated in the Swedish outbreak.

The bean sprouts implicated in the English outbreak were reportedly manufactured using a process common to most producers of this type of product. Although it is reported that bean sprout production involved washing the mung beans in water containing hypochlorite (7500 ppm), it is not clear whether this related to the beans implicated in this outbreak or whether this was a generic comment about bean sprout manufacturing at the time. After washing, the mung beans were rinsed prior to soaking in germination rooms (*c.* 27°C) for 24 hours. Growth of the sprouts was continued for a further 4–7 days at *c.* 20°C in a humid, dark environment with water being supplied on a continuous basis. Once the sprouts were of the desired size they were harvested and agitated in chlorinated water (100 ppm) to separate the sprouts from the seed coat and then packed by hand into large polythene bags or small film-covered punnets (O'Mahony *et al.*, 1990).

Another outbreak of salmonellosis occurred in England in June 1989, this time associated with mustard cress (Joce *et al.*, 1990). Thirty-one cases of *S.* Goldcoast infection were identified in the community and 14 of the 25 persons subsequently interviewed remembered eating cress in the week before their illness. *Salmonella* Goldcoast was not isolated from any product, seed, environmental or water sample taken as part of the investigation, but it had been isolated in a routine sample of cress taken by the factory at the end of May 1989. The source of the organism was not established although it was reported that the mains water supply used for the cress had been interrupted at the end of May 1989 necessitating the use of water stored in an uncovered reserve storage tank (Joce *et al.*, 1990). Although the tank was reportedly subjected to routine manual chlorination, analysis of the water in the subsequent investigation found coliforms and *E. coli* that, after effective cleaning and disinfection were eradicated. The producer used three types of seed (mustard, rape and cress) all imported from the Netherlands. The manufacturing process involved soaking the seeds in hyper-chlorinated water followed by dispersal of seeds into punnets of peat. The seeds were germinated in warm, humid rooms for 48 hours and then moved into glasshouses for approximately three days, during which time water was supplied via sprays. The product was sold in the punnets with a shelf life of approximately four days.

Sprouted salad vegetables have been implicated in a large number of outbreaks of foodborne illness, most frequently due to enteric bacterial pathogens such as *Salmonella* and *E. coli* O157. Alfalfa sprouts were identified as the source of *Salmonella* in a large multi-national outbreak of *S.* Newport infections in North America affecting 133 people (Van Beneden *et al.*, 1999) (Table 2.1) and more than 45 cases of *S.* Enteritidis infection linked to the consumption of raw mung bean sprouts were reported in California, USA (Anon, 2000f). Contaminated alfalfa seed, distributed to multiple growers across North America was believed to be responsible for the former widespread, protracted outbreak (Van Beneden *et al.*, 1999). This and further outbreaks have been reviewed by Taormina *et al.* (1999).

As described in the outbreaks discussed, sprouted salad vegetables are grown from seed under warm, humid conditions to encourage the rapid germination and development of the edible sprout/plant destined to become the finished product. However, these same conditions are also ideal for the proliferation of microorganisms.

Microbial contaminants present in/on the raw material seed or introduced to the growing environment from people, the environment or water, have the potential to grow to very high numbers during the sprouting process. Jaquette *et al.* (1996) demonstrated the growth potential of *S.* Stanley in a typical sprouting process. *Salmonella* Stanley was artificially inoculated and dried onto alfalfa seed to achieve a level of 3.29 log cfu/g dry seed. Initial levels increased to 5.97 log cfu/g during the first 24 hours germination stage, followed by an increase to 7.08 log cfu/g during the subsequent initial 18 hours of sprouting. Andrews *et al.* (1982) studied the growth of *Salmonella* on mung bean and alfalfa seeds during germination over a three to four day period. They reported increases ranging from three to nearly five log units (Table 2.2) over this timescale.

Most outbreaks of illness due to these products have implicated the seed as the source of the pathogen. Due to the nature of the sprouting process, contamination of the seed will invariably result in contamination of the final product and, therefore, seed quality is paramount to the safety of these products. Seed for mung bean sprouts, cress or alfalfa are grown as large-scale agricultural crops. The seed is often purchased on the open market and destined to be used for a wide variety of purposes including animal feed, oil extraction or for germination and use as salad vegetables. However, the agricultural conditions used to grow and harvest the seeds do not differ to reflect the differing risks associated with the end use of the material.

Table 2.2 Growth of *Salmonella* during the germination of mung bean and alfalfa seeds, adapted from Andrews *et al.* (1982)

Bean/seed	Serotypes	Level of *Salmonella* (log cfu/g)*		Increase during germination (log cfu/g)
		Inoculated to seed	Recovered from sprout	
Mung	*S.* Anatum and	0.41	5.23	4.82
	S. Montevideo	2.52	6.75	4.23
Alfalfa	*S.* Eimsbuettel	−0.16	3.15	3.31
	and *S.* Poona	2.18†	5.88†	3.7

* Mean of three replicates.
† One sample only.

Seeds are sourced from countries throughout the world and it is difficult to ensure appropriate agricultural controls are in place for these commodities. The use of animal wastes on the land, conditions of waste storage or treatment together with the quality of irrigation water and the harvesting and storage conditions of the subsequent seed crop will all influence seed contamination by enteric pathogens. Clearly, it is preferable if the seed can be sourced from known, reputable supply points that adhere to high standards of agricultural practice in crop growing and harvesting, and seed storage. Supported by routine supplier inspection in a supplier quality assurance programme, these are probably the best means of assuring seed of high quality and safety. However, this approach is only possible for some very large producers of sprouted seeds and is beyond the means of medium and small producers who generally buy from the open market. It is not clear whether any of the producers implicated in these outbreaks had extensive raw material quality assurance programmes.

Once received by the sprout producer, seed batches should be subjected to routine microbiological monitoring for specific pathogens like *Salmonella* and indicators of faecal contamination such as *E. coli*. In addition to testing each batch of dry seed, it may be useful to supplement this testing with trial germination under laboratory conditions using non-chlorinated sterile water prior to testing for pathogens and indicator organisms. In this way it is possible to assess the quality of the batch, under worst-case conditions, prior to use of the seed.

The seed should be subject to some form of microbial reduction process prior to germination. This is most commonly a pre-germination wash in hyper-chlorinated water to decontaminate the outside of the seed. Levels in excess of 100 ppm free chlorine are routinely used but levels up to 20 000 ppm calcium hypochlorite have been suggested as treatment processes for the decontamination of seeds (Food and Drug Administration, 1999a).

Soaking in chlorinated water will reduce bacterial contamination but may not entirely eliminate it. Studies by Jaquette *et al*. (1996) demonstrated that *S*. Stanley, artificially inoculated and dried onto alfalfa seeds, decreased from 339 cfu/g seed to 8 cfu/g and 37 cfu/g seed after dipping in a solution containing 1010 ppm chlorine for 5 and 10 minutes, respectively (Table 2.3). In a similar study using a combination of five *Salmonella* serotypes, Beuchat (1997) demonstrated a greater than 3 log reduction on alfalfa seeds when soaked in calcium and sodium hypochlorite solutions (1800 ppm and 2000 ppm, respectively) or hydrogen peroxide (6%) for 10 minutes (Table 2.4). However, even though levels reduced to less than 1 cfu/4 g of seed, treated seeds were positive in enrichment tests for *Salmonella*, demonstrating that the treatment had reduced but not eliminated the organism. It should be noted that bean sprout producers at the time of the *S*. Saintpaul outbreak in the UK were

Table 2.3 Effect of a chlorine dip on the survival of *Salmonella* Stanley on alfalfa seeds, adapted from Jaquette *et al*. (1996)

Inoculum (cfu/g seed)	Chlorine concentration (ppm)	Dip time (minutes)	Level on seed after treatment (cfu/g)*	Estimated decrease in comparison to initial inoculum (log cfu/g)
339	0	5	248	0.14
		10	387	−0.06
	100	5	128	0.42
		10	197	0.24
	290	5	59	0.76
		10	99	0.53
	480	5	48	0.85
		10	64	0.72
	1010	5	8	1.63
		10	37	0.96
65	2040	5	0	> 1.8
		10	0	> 1.8
	3990	5	0	> 1.8
		10	0	> 1.8

*Average of three samples.

Table 2.4 Effect of several decontamination treatments on the survival of *Salmonella* on alfalfa seeds, adapted from Beuchat (1997)

Treatment*	Concentration used	Level of *Salmonella* recovered (cfu/g seed)	Estimated decrease in comparison to control (log cfu/g)†	Germination of seed (%)‡
Sodium	0	7944	NA	92
hypochlorite	200	331	1.38	89
(ppm)	1100	7	3.05	90
	2000	< 1	> 3.9	91
Calcium	0	8720	NA	92
hypochlorite	160	240	1.56	89
(ppm)	900	16	2.74	94
	1800	< 1	> 3.9	91
Hydrogen	0	3710	NA	91
peroxide (%)	1	148	1.40	93
	4	34	2.04	89
	10	< 1	> 3.6	92
Ethanol (%)	0	5750	NA	92
	80	3	3.28	89

*Treatment time 10 minutes.
† Enrichment cultures of peptone water used to wash seeds after treatment were found to be positive for *Salmonella*.
‡ Germination of seeds after treatment for 10 minutes.
NA = not applicable.

reportedly soaking seeds in hyper-chlorinated water at levels of up to 7500 ppm chlorine (O'Mahony *et al.*, 1990). However, in an outbreak of salmonellosis caused by *S.* Mbandaka in 1999 in the USA where alfalfa seeds were distributed from one lot to six sprout producers, only the two processors that did not treat the seed with 2000 to 20 000 ppm calcium hypochlorite were implicated in the outbreak (Weissinger and Beuchat, 2000).

A more detailed evaluation of a wide range of decontamination systems for alfalfa seeds was conducted by Weissinger and Beuchat (2000). Reductions in a six-strain cocktail of *Salmonella* serotypes ranged from log 2.0–3.2 cfu/g after exposure of contaminated seeds for 10 minutes to the following: 20 000 ppm free chlorine (calcium hypochlorite), 5% trisodium phosphate, 8% hydrogen peroxide, 1% calcium hydroxide, 1% calcinated calcium, 5% lactic acid or 5% citric acid. Lower reductions were achieved using 1060 ppm Tsunami or Vortex (commercial treatment processes),

1200 ppm acidified sodium hypochlorite or 5% acetic acid. The viability of the seeds was also adversely affected by many of the treatment processes.

The use of alternative procedures for seed decontamination such as pasteurisation or ozone treatment may be important approaches to investigate in order to maintain the long-term viability of sprouting seed manufacture. Therefore, in addition to the controls detailed already, it is also advocated that spent process water and finished product samples are examined for indicator and pathogenic bacteria as part of the monitoring programme in the production of these products.

Although the most likely source of *Salmonella* is the seed, a variety of other sources of contamination exist in the production environment. Water was implicated in one of the outbreaks and as this is extensively used in the production of these products, it is essential that such water is of a potable standard. Water used during sprout growth should be chlorinated (*c.* 1 ppm) as most growing sprouts can tolerate concentrations of chlorine sufficient to keep water storage tanks and pipes free from vegetative bacterial contaminants.

The environment and equipment must be of a suitable hygienic design and construction and it should be capable of being cleaned and disinfected appropriately. Personnel access to the germination rooms should be limited and controlled and appropriate hygiene precautions must be taken prior to entry into such areas.

In some cases, e.g. salad cress, the product is grown in punnets of peat and this material must be incorporated into a raw material quality assurance programme that should include regular microbiological monitoring.

Where the finished product is washed prior to packing, e.g. mung bean sprouts, this should be done using chlorinated water, although contaminants entering at this stage are much less likely to proliferate than those present before germination and growth of the seed. Where the product is hand packed, appropriate hygiene precautions such as the use of dedicated and regularly disinfected gloves should be considered.

These products are sold in bags or punnets, usually under refrigeration conditions with shelf lives of 3–5 days. In some cases they may be sold under ambient conditions from market stalls but usually deteriorate very quickly and soon become organoleptically unacceptable if not stored chilled or consumed quickly. Product instructions on the pack usually advocate the washing of these products prior to consumption but reductions achieved by washing in the home are likely to be limited and will certainly deliver no more than a one log cycle reduction in the surface bacterial load.

Due to the high number of serious outbreaks in the USA, the Food and Drug Administration issued advice to consumers to avoid the consumption of raw vegetable sprouts in order to reduce the risk of foodborne illness (Food and Drug Administration, 1999b).

In the UK, following the outbreak due to bean sprouts in 1988, a code of practice was written to indicate recommended best practice for the safe manufacture of these products (Brown and Oscroft, 1989).

Sprouted salad vegetables are a high-risk product group subject to contamination and growth of enteric pathogens. Complete assurance of safety is not attainable at any one stage in the process, but applying strict controls from seed production through to the finished product can enhance the safety of these products. It is incumbent on the industry to employ adequate controls throughout the production process to limit outbreaks associated with these products.

SALAMI: ENGLAND

In January 1988, a number of cases of salmonellosis caused by identical *Salmonella* isolates were reported to the Communicable Disease Surveillance Centre (CDSC) in England. The infections were distributed throughout England and a full epidemiological investigation was initiated (Cowden *et al.*, 1989). This investigation revealed a total of 101 cases in England caused by the outbreak strain. Eighty-five cases were interviewed of which 72 were primary cases and 13 secondary cases. Forty-six of those interviewed were male and 39 female with the ages of those affected ranging from seven months to 78 years (median age six years). Eighty-one of the 85 cases reported suffering diarrhoea, 35 with blood in their stools. Thirty-eight individuals experienced vomiting and 71 fever. The symptoms lasted from one to 30 days (median seven days) and 19 patients were admitted to hospital; there were no deaths. Two patients suffered complications; one 14-year-old girl was diagnosed with appendicitis and had an appendicectomy and a 16-year-old boy developed acute ulcerative colitis (Cowden *et al.*, 1989)

The food implicated in this outbreak was a 'snack' salami (Table 2.5) consumed by 68 of the 72 primary cases (Cowden *et al.*, 1989). The products were small sized sticks of salami, sold in individual pre-packs, made in a factory in Bavaria and imported into the UK by a major food manufacturer.

The outbreak strain was *Salmonella* Typhimurium DT124 which was isolated from affected individuals and the implicated salami sticks. A total of 77 salami sticks from 13 different batches (11 on sale and two new batches) were examined for *Salmonella* spp. Five out of 10 sticks from one batch of salami were contaminated with *S.* Typhimurium DT124, which was also isolated from one of the new batches of product. An investigation by the manufacturer and importer found contamination to be restricted to five consecutive batches of salami sticks for which a single meat component was used in common. An investigation by the Bavarian health authorities found the standards of hygiene in place at the factory to be acceptable and no *Salmonella* could be found in over 200 samples taken from raw materials, equipment and the environment. Faecal samples from all of the factory workers also failed to reveal any carriage of *Salmonella* spp.

The salami sticks were manufactured from a mixture of ingredients including pork fat (50%), pork, beef, spices and curing salt. The fat was present as fairly large, discrete pieces distributed throughout the mix. No

Table 2.5   Outbreak overview: salami

Product type:	Raw, fermented meat – 'snack' salami
Year:	1988
Country:	England
Levels:	Not reported
Organism:	*Salmonella* Typhimurium DT124
Cases:	101 (no deaths)

Possible reasons
(i)   Presence of high levels of *Salmonella* in the raw meat (pork and beef) or other ingredients
(ii) Inadequate fermentation due to lack of added starter culture
(iii) Drying time too short to significantly reduce the initial levels of *Salmonella*
(iv) No other microbial reduction stage used, e.g. pasteurisation
(v)   Refrigeration of packed product aiding survival of *Salmonella*
(vi) Consumption of product by vulnerable groups

Control options*
(i)   Auditing of raw meat supply source supported by active monitoring of the microbiological quality of incoming batches of meat in a supplier quality assurance programme
(ii) Use of starter cultures and strict adherence to defined fermentation process requirements, i.e. pH decrease or acidity increase
(iii) Validation of drying time to ensure appropriate reduction in pathogens, e.g. 6 log reduction
(iv) Pasteurisation of product to provide enhanced safety for vulnerable groups

* Suggested controls are for guidance only. It is recommended that a full hazard analysis be carried out for every process and product to identify where controls should be implemented to minimise the hazard from *Salmonella*.

starter cultures were used to aid fermentation and due to the very small size of the product (final dimensions; 20 cm long, 1 cm diameter and 25 g weight), it was dried for only a short period of time (six days).

The finished product was placed in foil packages which were flushed with nitrogen, sealed and distributed at ambient temperature with a shelf life of six months. The product was distributed throughout Europe and sold under ambient conditions although it was common practice in the UK for these products to be sold from chill cabinets under refrigeration conditions, usually < 8°C.

Other fermented meat products have been implicated in outbreaks of salmonellosis. In 1995, 26 cases of *S.* Typhimurium infection were attributed to contaminated Lebanon bologna, a semi-dry fermented sausage (Sauer *et al.*, 1997). The product was made using beef that was ground with salt (3.3%), potassium nitrate (156 ppm) and stored refrigerated in tubs (7°C) for 10-14 days to encourage the growth of lactic microflora. The beef was transferred to a second plant where it was reground, stuffed into casings and smoked in a temperature controlled room to reach a temperature of 43°C over 52-72 hours. The process operated on the principle that as the meat mix gradually increased in temperature, fermentation would occur decreasing the pH to *c.* 4.5. Elevating the product temperature to 43°C over 52-72 hours also results in moisture loss and a decrease in water activity. No further drying process took place. The final product had a pH between 4.4 and 4.7, water activity of 0.95-0.97 and a salt content of 3.0-3.25%. It is reported that *S.* Typhimurium may have survived this process if present in the raw meat at sufficiently high numbers, i.e. $> 10^4$/g (Sauer *et al.*, 1997). The absence of a full drying process, a pasteurisation stage or other process combination capable of sufficiently reducing *Salmonella* are clearly factors that may have contributed to this outbreak.

Another large salmonellosis outbreak implicating a salami product caused by *S.* Typhimurium PT193 and involving 83 individuals in northern Italy in 1995 was believed to be due to a reduction in the normal salami process ripening time. The product was released onto the market after only 25 days ripening time (Pontello *et al.*, 1998), when the ripening time should have been 45 days.

The cause of the UK 'snack' salami outbreak is not entirely clear but as it was confined to batches of product manufactured from a common raw meat material, the starting point for this outbreak is most likely to have been *Salmonella* contamination of the raw meat. Raw meat is known to be contaminated occasionally with *Salmonella*, *S.* Typhimurium being one of the more common serotypes found in pork and beef. As *Salmonella* is likely to have been present on the raw meat on previous occasions, it is interesting to note that this product does not appear to have been implicated in any previous outbreaks of salmonellosis. Therefore, it is probable that the production process was generally able to reduce any initial levels of *Salmonella,* present as an occasional contaminant, to those not causing illness. Indeed, it was reported that over 26 million of these salami sticks were being imported annually into the UK and it is unlikely that such volumes of salami production would not have previously encountered some raw material meat batches contaminated with *Salmonella* even at a low incidence and level.

The outbreak is therefore most likely to have arisen due to one of the following factors:

(a) high levels of *Salmonella* contamination occurring in the single common meat component used for the implicated salami sticks
(b) inadequate fermentation and/or drying conditions leading to a lower than normal inhibitory effect on *Salmonella*
(c) the presence of a strain of *Salmonella* particularly resistant to the antimicrobial effects of the salami production process
(d) preservation of surviving *Salmonella* in the contaminated salami due to chilled storage of the product in the UK.

Indeed, it is possible and perhaps most likely that a combination of these factors resulted in *Salmonella* surviving the process to cause this outbreak.

The levels of contamination of the raw materials, including the meat, were not known but it should be anticipated that *Salmonella* will be occasionally present in the raw meat ingredients. Any salami manufacturing process should be designed to achieve a significant reduction in levels of this hazardous enteric organism.

It is clear that the production process used for the 'snack' salami was considerably shorter in time than that of standard salami manufacturing processes. Most European production processes employ fermentation conditions of 20–30°C for 2–5 days prior to extended drying periods of several weeks at < 15°C. The process for the 'snack' salami involved a drying stage lasting only six days due to the very small diameter of the product. The long drying times of many fermented meat products allow the acids produced during fermentation, together with other inhibitory additives such as nitrite and the decreasing water activity, to combine to effect significant reductions in levels of vegetative bacterial contaminants that may be present. Reduced drying times may result in greater survival of enteric pathogens and many of the reported salami-related outbreaks of salmonellosis implicate products with short or shortened drying times.

In addition, this process did not employ the use of starter culture bacteria and presumably relied on the fermenting action of lactic acid bacteria naturally present in the raw materials and/or added acidity agents. Starter culture bacteria do not, in themselves, achieve a reduction in the level of pathogens; the inhibitory agents that they produce together with those added to the product contribute to pathogen reduction. In addition, added starter bacteria can make the fermentation process and concomitant production of inhibitory agents more rapid, consistent and reliable.

Investigations by the manufacturer found the contaminated batches of product to have a higher pH and glucose concentration and a lower acidity content than normal. This may reflect either poor lactic bacterial activity or insufficient time allowed for the fermentation stage.

More detailed studies on the effect of manufacturing conditions on the survival of *S.* Typhimurium DT124 in the 'snack' salami product made with artificially contaminated raw material, failed to demonstrate any significant relationship between its survival and the use of starter cultures or glucono-δ-lactone. This was despite the fact that the pH reduced to < 5.2 during manufacture. However, marked differences in survival of *Salmonella* were found when the storage temperature of the salami was varied. A five log reduction of *S.* Typhimurium DT124 was achieved after 30 days storage at 15°C, 20°C and 25°C in the artificially contaminated salami sticks. At 10°C, the reduction was only three log units and at 5°C initial levels of contamination were only reduced by two log units. Similar reductions were found to occur with naturally contaminated salami sticks incubated at 20°C and 5°C (Cowden *et al.*, 1989). It is not reported how much reduction was achieved after the six-day process drying period following fermentation.

As the salami sticks were generally refrigerated in the UK, it is possible to suggest the following scenario to explain the outbreak. Firstly, the overall process time was extremely short, which is likely to result in only a small reduction of any contaminating enteric pathogens present. As bacterial pathogen contamination of raw meat, when present, is not usually at high levels, on most occasions the process would render the product safe to consume. However, for contamination loads slightly higher than normal, the process may not achieve adequate reduction at the manufacturing stage and levels sufficient to cause illness could have remained in the product at the point of packing. In most other countries where the product was distributed, it would have been sold and stored under ambient conditions where further reduction in *Salmonella* levels would have occurred, as demonstrated by the results of the challenge test trial. However, in the UK, the salami sticks would have been chilled, resulting in less reduction in the levels of any contaminating *Salmonella* spp. and leaving sufficient numbers to cause illness after consumption.

It can be seen clearly from this outbreak that raw material quality is critical to product safety and a supplier quality assurance programme should include regular monitoring of the raw material supply, meat in particular. This should include auditing of the supplying abattoir and meat cutting plant together with routine testing of the meat at the point of receipt for indicator microorganisms such as *E. coli* and specific pathogens such as *Salmonella*. In most cases, tests for indicator bacteria are of more practical use as they are usually present more frequently and their presence in high numbers can provide a warning of inadequate standards of hygiene during the slaughter and meat processing operations.

It should have been anticipated that any traditional process such as salami manufacture, when reduced in time to the extent apparent from the investigation of this outbreak, may allow insufficient time for the antimicrobial factors to effect an adequate reduction in the levels of contaminating bacterial pathogens. This should always be tested using challenge test studies to verify the ability of a process to effectively reduce anticipated levels of undesirable bacterial contamination.

It is interesting to note that the majority of cases in this outbreak were in young children due to the appeal of these products to this consumer group, presumably as snacks and for lunches and between meals. It is incumbent on the industry, when developing products such as these, to consider the likely consumer groups and to build in appropriate safety margins where the target groups may be more vulnerable to gastrointestinal infection. Such a safety margin could have been achieved by a longer drying process or, indeed, by pasteurisation of the product after processing. In fact, many products of this nature, i.e. 'snack' type salamis, are now pasteurised in-pack after processing. It is important to remember that 'normal' pasteurisation processes applied to products like these may not be very effective for destroying *Salmonella* spp. because of the protective effect of low water activity (see Chapter 3). Higher temperatures or longer times may be needed to achieve sufficient destruction of vegetative bacterial pathogens and challenge test studies to determine an effective heat process should be conducted to verify adequate destruction of the target microorganism.

ORANGE JUICE: USA

In June 1995, the New Jersey Health Department, USA was alerted by reports of seven cases of infection caused by an unusual *Salmonella* serotype. All individuals had recently returned from a vacation in Orlando, Florida having all visited a large theme park (Cook *et al.*, 1998). A subsequent epidemiological investigation identified a total of 52 confirmed and 10 probable cases of infection with the same *Salmonella* serotype. Based on the predicted under-reporting of infection, it was estimated that the outbreak probably affected between 1240 and 6200 individuals (Cook *et al.*, 1998).

The ages of those affected ranged from one to 63 years (median 10 years) and the symptoms included diarrhoea (100%), abdominal cramps (97%), fever (97%), headache (73%), bloody stool (71%) and vomiting (66%). Illness lasted from three to 23 days (median seven days) and seven people were hospitalised. No deaths were reported.

The implicated product was fresh pressed orange juice (Table 2.6) and 31 of the 32 persons in the case control study recalled consuming this product. The orange juice was manufactured by a local juice producer and was unpasteurised.

The outbreak strain was identified as *Salmonella* Hartford (*Salmonella* serogroup C_1 - Kauffmann–White Scheme), which was isolated from 58 stool samples, one urine sample, two blood samples and one abscess sample taken from affected individuals. *Salmonella* Hartford and *S.* Gaminara were both isolated from the stool of one patient.

Fruit was supplied to the manufacturer from several local orchards. Prior to use, the fruit was passed through a phosphoric acid wash in pallets on a conveyor (Cook *et al.*, 1998). The fruit was then rinsed with water, sliced in half and pressed. The juice was then chilled and bottled. No preservatives were added to the product and the pH was reported to be pH 4.1–4.5 (mean 4.3).

Salmonella Gaminara was isolated from 10 out of 12 samples of orange juice (2–4 cfu/100 ml) produced at the implicated factory between May and July 1995 and this was taken as indicative of ongoing contamination attributed to inadequately sanitised equipment (Cook *et al.*, 1998).

Another salmonellosis outbreak was detected in June 1999 when two State health departments in the USA (Washington and Oregon) recorded

Table 2.6   Outbreak overview: orange juice

Product type:	Unpasteurised fresh pressed orange juice
Year:	1995, 1999, 1999
Country:	USA, USA/Canada, Australia
Levels:	Not reported
Organism:	*S.* Hartford, *S.* Muenchen, *S.* Typhimurium PT135a
Cases:	62, 298, > 400 (no deaths)

Possible reasons
(i)   Use of fallen fruit possibly contaminated by soil/animal faeces
(ii) Inadequate decontamination of fruit prior to processing
(iii) Possible internalisation of pathogens in fruit
(iv) Introduction of *Salmonella* due to inadequate biosecurity allowing birds, rodents and amphibians into the processing environment
(v)   Inadequate cleaning/sanitisation of fruit transport containers and conveyors, washing, pressing and filling line equipment

Control options*
(i)   Good agricultural practice by application of only properly treated waste or artificial fertilisers to land and prevention of access to orchards by grazing animals
(ii) Use of tree picked fruit only with no fallen fruit
(iii) Incorporation of effective sanitisation of fruit prior to pressing, e.g. use of chlorinated water
(iv) Attention to detail in manual cleaning and sanitisation of the plant
(v)   Biosecurity to prevent entry of insects, rodents, birds and amphibians
(vi) Pasteurisation of fruit juice or application of other *Salmonella* destruction process

*Suggested controls are for guidance only. It is recommended that a full hazard analysis be carried out for every process and product to identify where controls should be implemented to minimise the hazard from *Salmonella*.

clusters of diarrhoeal illness associated with one serotype of *Salmonella*. In the subsequent investigation, a total of 298 cases of infection caused by the same serotype were reported across 15 States in the USA and two Canadian Provinces (Boase *et al.*, 1999).

The outbreak was first suspected in Washington on 19 June 1999 when three cases of infection were identified, all of whom had consumed the same type of drink from different outlets of the same restaurant chain. Eighty-five persons in Washington were subsequently identified as having onset of illness between 10 June and 30 June, 1999. Of 79 patients from whom information was available, their ages ranged from nine months to 95 years (median 27 years) and 51% were male. Symptoms included diarrhoea (94%), bloody diarrhoea (43%) and fever (75%). Eight of the patients were hospitalised and none died.

In Oregon, an infection with the same *Salmonella* serotype prompted a case control study which found four ill persons who had attended the same buffet in Portland. Illness was significantly associated with the consumption of a drink produced by the same company who manufactured the drink implicated in the Washington State outbreak. Fifty-seven persons, of whom 54% were female, were subsequently identified as having suffered illness caused by *Salmonella* Muenchen, and the median age of those affected was 36 years. Similar symptoms to those described in the Washington outbreak were reported; diarrhoea (100%), bloody diarrhoea (59%), fever (89%), abdominal cramps (85%) and chills (82%). None of the patients died although seven were hospitalised (Boase *et al.*, 1999).

Subsequently, cases caused by the same *Salmonella* serotype were reported in 13 other States in the USA and two further Provinces of Canada.

The outbreak strain was *Salmonella* Muenchen (serogroup C_2) (Table 2.6) and the implicated drink was unpasteurised fresh orange juice produced by a single major manufacturer. Under seven different brand names, the manufacturer supplied a large number of retail and catering outlets across 10 USA States and two Canadian Provinces. The product was also sold in other States through secondary distribution systems and it was also distributed in a frozen form for use in restaurants and institutions.

A total of 67 out of 85 persons in the Washington State outbreak recalled either consuming unpasteurised orange juice made by the implicated manufacturer or eating at an establishment where the manufacturer's products were sold. Forty-four of the 57 cases in the Oregon outbreak reported drinking unpasteurised orange juice.

A recall of the unpasteurised orange juice produced by the implicated manufacturer was initiated on 25 June 1999 but it was not until 28 June that the outbreak strain was first isolated from an unopened container of the product. Subsequently, *S.* Muenchen was isolated from samples taken from a fruit juice blender and a juice dispenser in a Washington branch of the restaurant chain selling the implicated juice manufacturer's products. Samples of fruit juice taken from the orange juice manufacturing plant were found to be contaminated with *S.* Javiana, *S.* Gaminara, *S.* Hidalgo, *S.* Alamo and *S.* Muenchen (Boase *et al.*, 1999).

In July and November 1999, the US Food and Drug Administration issued two further health warnings relating to these products. The first was due to continued reports of illness associated with consumption of the products (Food and Drug Administration, 1999c) and the second due to the detection by the manufacturer of more lots of contaminated product (Food and Drug Administration, 1999d).

In South Australia in February/March 1999, more than 400 people were affected in an outbreak of food poisoning caused by *Salmonella* Typhimurium PT135a also attributed to unpasteurised orange juice (Anon, 1999d).

Fruit juice is being implicated in a growing number of foodborne outbreaks of infection, most commonly due to enteric pathogens. Some of these outbreaks have involved *E. coli* O157 and Bell and Kyriakides (1998) have reviewed some of these outbreaks, together with factors contributing to them.

Fruit juice is manufactured from fresh fruit collected from orchards and transported to the processor. Fruit may be stored for extended periods prior to use. The fruit is normally washed before pressing and the juice is collected via extraction pipes into receiving vessels. Prior to filling into catering size or smaller volume consumer packs, the juice may have pulp removed, although many products are produced with fruit pieces remaining as a sign of authenticity, freshness and quality. Preservatives may be added to prevent the growth of spoilage yeasts but this is not usual practice as this compromises the premium quality of fresh pressed juice. In the absence of added preservatives, fresh juice is usually given a shelf life of between 6 and 10 days if stored at low temperature ($<8°C$), although the juice in the Orlando outbreak was reported to have a shelf life of 12–17 days (Cook *et al.*, 1998).

The reasons for the Washington/Oregon outbreak are not entirely clear and therefore it can only be surmised how such an outbreak could occur. However, information relating to the cause of the Orlando outbreak is much more revealing. Firstly, it was reported that oranges were supplied to the processor from a number of local orchards. Fruit was hand-picked from the trees, dropped to the ground, collected in large bins and then delivered to the processing plant within 24 hours, under ambient conditions. Microbiological examination of oranges and soil swabs from the two orchards supplying most of the fruit found no *Salmonella* species in 95 orange samples but *S.* Braenderup and *S.* Muenchen were isolated from two environmental swabs taken from the soil surrounding trees in two tracts of one orange grove.

Historical outbreaks of illness implicating fruit juice and associated with enteric pathogens have commonly highlighted deficiencies in fruit collection practices as being contributory factors. Fruit used for juicing is often of a lower grade than fresh whole-fruit as the physical state of the fruit for juicing is of little importance to the customer. Top grade fruit is usually selected for sale as fresh whole-fruit (table-fruit) because the visual quality is a key determinant in the purchase decision. As a consequence there is a real possibility that the poorer quality fruit used for fruit juice production may include a significant proportion of fallen fruit picked from the ground because of the desire to avoid wastage. Clearly, in this outbreak, it appears that fruit was actually dropped to the ground after picking.

Fallen fruit has been implicated as a vector of microbial pathogens in previous outbreaks of foodborne illness (Bell and Kyriakides, 1998). Fruit becomes contaminated from soil onto which animal wastes may have been distributed either by the farmer for the purposes of soil fertilisation and conditioning or by free roaming animals. The use of fallen fruit is undesirable principally because of the potential to become contaminated with enteric pathogens but also, because they are more likely to be prone to colonisation by yeasts and moulds through bruising and skin damage caused by impact with the ground. Product quality can then be compromised through the introduction and growth of this spoilage microflora.

Enteric pathogens are known to be present in fresh animal slurries and such material should not be applied to land without appropriate consideration of the potential for pathogen transfer to crops. The risk of contamination from such sources can be reduced by the application of good agricultural practices in the use of fertilisers on soil. Use of properly composted manure can significantly reduce the pathogen burden on the soil due to the high temperature (60°C or more) achieved during the composting process. The use of artificial fertilisers or composted manure, as described, should be important considerations in fruit growing.

Use of dropped fruit should be minimised and most certainly, fruit should not be dropped deliberately after picking.

Once received at the juice manufacturing plant the fruit are usually washed prior to juice extraction. Washing has historically been conducted to remove dust and soil and other foreign material from the fruit prior to juice extraction so as to avoid physical contamination of the juice. In recent years washing has been the focus of much attention in trying to achieve a reduction in the levels of microflora on the fruit itself. Washing of oranges can be carried out more vigorously than other fruit because of its thick skin. In the Orlando outbreak, it is reported that the oranges were washed in phosphoric acid prior to rinsing with water (Cook *et al.*, 1998) and Parish (1998) indicates the use of an acid-anionic cleaning solution at a level of 200 ppm.

Fruit are usually washed using a brush washer to assist the removal of physical debris. Some processors wash fruit in tanks of potable water and whilst this can reduce the level of microorganisms on the surface of fruit by about 1 log unit, it is important to note that, as the organisms are washed into the water, they can be readily spread to other fruit. Indeed, as the organisms build up in the water, successive batches can leave the washer with higher levels of microbial contamination than they had upon entering it! In addition, localised contamination with pathogens such as *Salmonella* on just a few fruit can readily be spread throughout the batch via the water system. Chlorination or the use of other sanitisers in the water can help reduce cross-contamination as the chlorine can inactivate the organisms in the water. Washing in chlorinated water is usually the method of choice for fruit washing systems but it is important to ensure that the quantity of free chlorine is maintained at levels sufficient to kill pathogens washed off the fruit and into the water. Reductions in vegetative bacteria of 1–2 log units can be achieved by effective chlorinated washing systems but it is important to remember that chlorine is readily inactivated by organic debris and soil. Therefore, systems employing a prewash to remove debris and soil followed by a chlorinated wash or rinse are likely to facilitate better microbial reductions than chlorinated washing alone.

Although washing fruit is regarded as a useful intervention step for removal of contaminating pathogens from fruit surfaces, recent evidence demonstrating the internalisation of pathogens shows the vulnerability of total reliance on such controls. Research has indicated that some pathogens could become internalised in the fruit through scar tissue and this could protect organisms from the effect of surface cleaning and disinfection (Merker *et al.*, 1999). However, washing is still advocated as an important intervention measure, if not entirely effective.

Once washed, the fruit implicated in the Orlando outbreak was cut in half and pressed. Larger scale fruit processors use systems that make small cuts around the base of the fruit prior to pressing. The juice is collected into vessels via extraction pipes situated below the extractor. The extraction of juice from oranges differs significantly from the process applied to apples which are pressed or milled together. It is argued that as less contact is made between the skin and juice in the extraction of orange juice, this may reduce the potential for transfer of microbial contaminants, although it should be remembered that significant contact still occurs between the juice and the skin surface.

The fruit is often cited as an important source of pathogens in outbreaks of enteric infection implicating fruit juice, but it should be noted that sources other than the fruit were implicated as contributory factors in the Orlando outbreak. It was reported that contamination could have been introduced by the entry of animals into the processing area and that inadequately sanitised equipment may have prolonged the outbreak (Parish, 1998). It was reported that the processing equipment was significantly exposed to external sources of contamination. The processing room was poorly sealed with cracks and holes clearly visible in the walls and ceiling (Cook *et al.*, 1998). Rodent and bird droppings were found in the processing room and it was reported that frogs were observed around processing equipment.

Samples taken around the processing environment during the subsequent investigation included dried frog faeces, four tree frogs and a toad. Readers will be comforted to know that the frogs were sampled for the presence of *Salmonella* by leaving overnight in peptone buffer prior to release, rather than by conventional blending! (Parish, 1998)

Salmonella Newport was isolated from tree frogs and *S.* Newport and *S.* Hartford were isolated from the toad, all of which were found outside the processing building (Cook *et al.*, 1998). In addition, the investigation found *S.* Saintpaul on the surface of unwashed oranges held in cold storage for several weeks.

In relation to process environment hygiene, *E. coli* was found at high levels on the orange conveyor belts prior to the washing stage and in large numbers (10^5/swab) from floor drains (Parish, 1998). To assess the efficacy of cleaning, a comparative study was conducted of the microbiological cleanliness of equipment and product before and after a more intensive cleaning and sanitisation regime was introduced that included a terminal spray of hypochlorite (200 ppm). After improvements to equipment cleaning and sanitisation procedures and inclusion of an additional hypochlorite wash of the fruit (200 ppm for 20 seconds), resulting juice samples had 96% fewer microorganisms than the cleanest previous commercial sample. In addition, seven out of eight previously produced commercial samples of orange juice contained *E. coli* levels between 4 and > 110 MPN/ml, but following the improved procedures, *E. coli* was not detected (< 0.3 MPN/ml) in samples of finished product.

As unpasteurised juice is not normally subject to any further processing that reduces the level of microbial contamination, many of the micro-organisms present on the fruit after washing will be present in the finished product. Historically, it was thought that fruit juice had such a low pH that vegetative pathogens would die out in the product. However, the now numerous outbreaks of illness recorded caused by *E. coli* O157 and *Salmonella* tend to suggest otherwise.

Research has shown that *Salmonella* serotypes can survive for several weeks in orange juice, although levels do reduce over time. Parish *et al.* (1997) determined the survival of several *Salmonella* serotypes in orange juice. To achieve a 6 log reduction in *Salmonella* serotypes, orange juice (pH 3.5) had to be stored at 4°C for 15–24 days. A similar reduction took 43–57 days when the orange juice was pH 4.1 (storage temperature 4°C). Although using apple juice, a study by Uljas and Ingham (1999) demonstrated that *S.* Typhimurium DT104 could be reduced by at least 5 log units at pH 3.3 after storage at 25°C for 12 hours or, at 35°C for 2 hours. These treatments did not achieve a 5 log reduction in *E. coli* O157. At pH 4.1, a 5 log reduction in *S.* Typhimurium DT104 was achieved by storage at 35°C for 6 hours in the presence of 0.1% sorbic acid or by a combination of storage at elevated temperature (25°C for 6 hours or 35°C for 4 hours) followed by a freeze/thaw cycle without sorbic acid (Uljas and Ingham, 1999).

In the USA, concerns relating to the potential health hazards associated with the consumption of unpasteurised fruit juices has led the Food and Drug Administration to require health related labelling on such juices. The health warning reads 'WARNING: This product has not been pasteurised and, therefore, may contain harmful bacteria that can cause serious illness in children, the elderly, and persons with weakened immune systems' (Food and Drug Administration, 1998). This statement may be left off the pack if the manufacturer applies 'a reduction that is equal to, or greater than, the criterion established for process controls by any final regulation requiring the application of Hazard Analysis and Critical Control Point (HACCP) principles to the processing of juice' (Food and Drug Administration, 1998), or if the processor applies a process capable of achieving a 5 log reduction in the relevant organism. In the past, it has been acceptable for compliance to be achieved by the application of washing treatments or other surface decontamination approaches. However, the recent research on internalisation of pathogens in fruit has led the Food and Drug Administration to consider applying the 5 log reduction standard to stages of the process after juice extraction (Food and Drug Administration, 1999e) (see Chapter 4).

Clearly, more research is needed to clarify the full extent of the dangers presented by internalisation of pathogens in fruit but, one fact is clear, if microbial pathogens are not introduced onto fruit in the first place, there would be much less of a health risk. The application of effective agricultural practices in the orchard, avoidance of fallen fruit, biosecurity of production facilities and effective cleaning and sanitisation of equipment and fruit would significantly diminish the danger of outbreaks of illness from consumption of these products without the need to resort to complex and costly pathogen reduction programmes in the finished product.

CHOCOLATE CONFECTIONERY: USA AND CANADA

Between December 1973 and February 1974, 80 cases of infection due to a rare *Salmonella* strain were reported by 23 States in the USA (Craven *et al.*, 1975). Similar findings were being reported across Canada with 39 additional cases identified in seven Canadian Provinces. A case control study identified a common food item as the cause of the outbreak. In total, over 200 cases of infection resulted from the outbreak strain (D'Aoust, 1977), 122 in the USA (Craven *et al.*, 1975) and 95 in Canada (D'Aoust *et al.*, 1975). No deaths were recorded.

The ages of those affected in the USA outbreak ranged from one day to 79 years (median three years). Twenty-seven of the 71 patients with symptoms were admitted to hospital with severe diarrhoeal illness and one person suffered severe intestinal bleeding requiring transfusions of eight units of blood (Craven *et al.*, 1975). Two, one-day-old babies were infected with the organism; the mother of one had suffered gastrointestinal illness four days before delivery and the mother of the second baby had consumed the implicated product three days before delivery. The outbreak strain was isolated from a stool of the first mother but not the second. The age range of individuals in the Canadian outbreak was similar (few days to 75 years) and symptoms suffered included nausea, vomiting, diarrhoea and fever (D'Aoust *et al.*, 1975).

The implicated product in both outbreaks was chocolate confectionery items manufactured by a single production facility in Canada. Fifty-three people in the USA outbreak had a history of consuming chocolate manufactured by the implicated producer; 46 had eaten Christmas-wrapped chocolate balls, seven ate a similar seasonal chocolate product and a further seven people contracted the illness through secondary spread from infected individuals. In the Canadian outbreak, 30% of those affected were reported to have consumed chocolate balls or similar items produced by the implicated manufacturer within three days of onset of symptoms (D'Aoust *et al.*, 1975).

The outbreak serotype was identified as *S.* Eastbourne (Table 2.7), which was isolated from chocolate products retrieved from the homes of some of the affected individuals. The estimated contamination level of the chocolate was 2.5 *S.* Eastbourne cells/g (Craven *et al.*, 1975). The Canadian authorities conducted a further investigation of retail chocolate from the implicated manufacturer testing a total of 575 samples of over 10 different chocolate items (D'Aoust *et al.*, 1975). In total 54.35 kg of product was examined for *Salmonella*; the isolation rate was 7.7% (44 out of 575

Table 2.7   Outbreak overview: chocolate confectionery

Product type:	Chocolate balls and other chocolate novelty confectionery	
Year:	1973/4	
Country:	USA/Canada	
Levels:	2–9 cells/chocolate ball	
Organism:	*Salmonella* Eastbourne	
Cases:	217 (no deaths)	

Possible reasons

(i)   Use of cocoa beans contaminated with *Salmonella*

(ii) Inadequate heat processing of the raw cocoa beans

(iii) Inadequate segregation of roasted beans from raw beans

(iv) Heat processing of subsequent chocolate mixture inadequate to destroy *Salmonella* due to the low water activity of the mixture

(v)   Consumption by vulnerable groups

Control options*

(i)   Application of good agricultural and hygienic processing controls during bean growth, harvesting, fermentation and drying

(ii) Application of effective roasting of beans to destroy *Salmonella*

(iii) Segregation of roasted beans from raw beans incorporating physical separation of beans, control of personnel and air-handling systems

(iv) Effective control of water used for chocolate tempering

(v)   Effective supplier quality assurance programme covering other ingredients

* Suggested controls are for guidance only. It is recommended that a full hazard analysis be carried out for every process and product to identify where controls should be implemented to minimise the hazard from *Salmonella*.

samples). Twenty-seven of the positive samples were chocolate balls, with an estimated level of 2–9 cells of *S.* Eastbourne per contaminated chocolate ball (D'Aoust *et al.*, 1975). Production dates of the confectionery found to contain *S.* Eastbourne ranged from May 1973 to September 1973.

A detailed investigation of the production plant found *S.* Eastbourne in samples taken from two cocoa bean processing rooms and the chocolate moulding plant. The positive isolates in the bean rooms were obtained from the bristles of hand-brooms, a bucket and the gear mechanism of a conveyor. In the moulding area, one of the isolates was recovered from a chocolate ball on the floor and the other from chocolate encrusted onto the inside lip of a tempering tank. An additional isolate was recovered from roasted cocoa beans that had been cleaned and were being air-cooled in a room prior to passing back into the room where raw beans were present.

Other chocolate confectionery-related outbreaks of salmonellosis have reinforced the safety risk in relation to chocolate. Gill *et al.* (1983) reported a *S.* Napoli outbreak in England and Wales due to consumption of contaminated bars of chocolate imported from Italy. Two hundred and forty-five cases of infection were reported, with a significant number of further cases being prevented by the prompt recognition of the outbreak and the subsequent public health warning and product recall. In total, about 32 tons, which was four-fifths of the entire batch of product, were recalled from sale. Estimates of the level of contamination indicated that *Salmonella* could have been present in numbers as low as 2 organisms/g of chocolate (Gill *et al.*, 1983). Although the source of *Salmonella* was not identified, potential sources of contamination were considered to include raw beans, untreated water used warm for keeping chocolate molten in storage vessels, and other ingredients such as dried milk, cocoa butter and chocolate crumb (Gill *et al.*, 1983).

Chocolate manufacture starts with the harvesting and fermentation of cocoa beans. Cocoa beans are sourced from a variety of countries, and hygienic practices associated with harvesting and subsequent fermentation vary considerably. Seed pods are broken open and the beans are removed together with some of the pulp residues that surround the beans in the pod. The beans are then fermented for several days by natural microflora which gain access to the beans from the soil, air, surface of the pod and the handling during removal from the pod. Fermentation assists the development of cocoa flavour and is followed by drying the beans, usually in the sun, to a very low moisture content (<8%). The beans are then sold to chocolate manufacturers who subject them to further processing.

Clearly, the cocoa beans are subject to environmental contamination during harvesting, fermentation and subsequent sun drying. With such open systems of production it is not possible to exclude the potential for *Salmonella* and other enteric pathogens to be present in the raw material. Application of good agricultural practices during harvesting and fermentation can reduce the incidence and level of microbial contamination but exclusion of insects and other vectors of contamination is difficult in a typical sun-drying process without the introduction of physical barriers that completely surround the material. It must therefore be anticipated that cocoa beans will, on occasion, be contaminated with enteric pathogens such as *Salmonella*.

The beans implicated in the USA/Canadian outbreak were sourced from a variety of countries including Ghana, Nigeria, Brazil and Ecuador and there were no positive isolations of *Salmonella* from any of the batches of beans remaining on the chocolate manufacturer's site.

Once received at the factory, in a raw bean room, the beans were cleaned by agitation to remove physical debris. They were then conveyed in a tube to a second room where they were subjected to roasting using hot dry air for half an hour, during which time the beans were brought to a temperature of 125°C. Roasted beans were then cooled in a hopper using room air and then conveyed back to the raw bean room where the shells were removed. The nibs (cotyledons) were then sent back into the second room to be ground to form cocoa liquor which was held at 65°C for 48 hours (D'Aoust *et al.*, 1975). The milk chocolate was made by blending salt, sugar, lecithin, vanillin and 'crumb' (a mixture of 85% dried whole milk, 5% cocoa powder and 10% sugar) with cocoa butter and cocoa liquor in 3000 kg batches. The chocolate was conveyed on open conveyors to roller refiners and then to conches (large mixers). The mixture was swirled gently in conches at temperatures of *c.* 60°C for periods of 5–24 hours to achieve the correct consistency. Due to the heating and cooling stages involved in the process, chocolate was frequently held at temperatures of *c.* 60°C for a total of only 6 hours. The resulting moisture content of the product was *c.* 0.5%. The chocolate was then held in storage tanks at 40°C in a molten state before being moulded into different shapes.

If microbial contamination is introduced into a chocolate manufacturing plant, it is very difficult to eradicate, as seen in this outbreak. Chocolate has an extremely low moisture level and associated low water activity (< 0.6) and thermal destruction of even vegetative bacterial contaminants is almost impossible to achieve during the conching process. The stage where most destruction can be achieved is during roasting of the beans. Roasting temperatures of 120°C or more can significantly reduce the level of microbial contamination of the cocoa beans and it is reported in this outbreak that the beans were brought to a temperature of 125°C during roasting. It follows that if roasting is a critical stage in decontaminating the bean, the roasted beans should be effectively segregated from the raw beans to prevent cross-contamination. The USA/Canadian outbreak investigation found *S.* Eastbourne in the raw bean processing area and it is evident that after roasting, the beans were transported back into this room for shelling and then back again into the roasting room for milling and refining.

Significant opportunity for cross-contamination exists if raw beans and roasted beans are stored or processed in the same environment. Large amounts of dust will be present that can be readily spread between the two materials via the air or by personnel. The finding of *S.* Eastbourne in a sample of roasted beans being air-cooled gives significant credence to the theory that cross-contamination from raw cocoa beans was the cause of the outbreak.

The principles of segregation of raw from cooked products should be applied to the processing of raw and roasted beans. This should include physical barriers between these materials, control of air sources for cooling of the roasted beans and control of personnel between the two areas.

The only other stage, apart from roasting, where temperature is significantly elevated is during the conching process. On the face of it, heating a product to 60°C for 5-24 hours would be expected to achieve a significant reduction in the levels of vegetative microorganisms, including *Salmonella*. However, microorganisms are protected against the effect of heat at low water activity levels and, as the chocolate has such a low amount of moisture (< 2%), the effect of the heat over this period is almost negligible. Studies carried out by Goepfert and Biggie (1968) demonstrated that in molten chocolate, *S.* Typhimurium had a *D* value at 70°C of 720-1050 minutes. Conching may achieve a reduction in levels of contaminating microorganisms, but the reduction will be very small indeed, and this outbreak demonstrated how contamination, entering at an early stage, could survive the processing employed in the manufacture of chocolate. Clearly, once introduced into the chocolate manufacturing process, contamination will be almost impossible to remove without closure of the factory and full cleaning, so the emphasis for control must be focused on eliminating *Salmonella* in the raw beans and subsequently preventing any cross-contamination.

The potential for the introduction of microbial contaminants at other stages of the chocolate making process should not be overlooked. Water used for keeping tanks and pipes warm could be a source of contamination and needs to be under suitable control. In addition, effective training of staff in hygienic practices is essential to prevent people from transferring contaminants from themselves or from external sources to the product.

Following the outbreak associated with chocolate balls, the manufacturer implemented a number of improvements to provide greater protection against the potential for *Salmonella* contamination. These included complete segregation of raw bean storage and roasting areas, dry roasting the beans at 125°C followed by moist heat treatment of the nibs and increased holding temperature of the chocolate liquor following the roasting stage and during the conching process (D'Aoust *et al.*, 1975).

It is interesting to note that the level of contamination required to cause infection was very low, probably < 100 cells (D'Aoust *et al.*, 1975). This is consistent with other outbreaks of salmonellosis involving food with a high fat content (Table 1.5). It is apparent that the high fat level gives some degree of protection to the organism in its passage through the stomach, allowing lower initial levels to survive and cause an infection. The consumption of chocolate by vulnerable groups such as young children, who represented a significant proportion of the outbreak cases, will also have been a significant factor in accounting for the apparent low infectious dose.

Chocolate manufacture is vulnerable to contamination by *Salmonella* and the application of a properly structured HACCP approach to its manufacture is essential to ensure that lessons learnt from outbreaks affecting previous generations are not re-lived at first hand by future ones.

INFANT DRIED MILK: ENGLAND

In December 1985, a large number of reports of isolations of one unusual *Salmonella* serotype were observed across England (Rowe *et al.*, 1987). A total of 76 people were reported to be infected with the outbreak strain, although in 12 cases, the implicated product could not be confirmed. Forty-eight cases were reported in infants. The predominant symptom was diarrhoea although the duration and severity was variable. Seven of the 48 infants were admitted to hospital due to the severity of their illnesses and one infant died. A milder illness was reported in older children and adults and of 28 non-infant cases, 12 had no symptoms.

The product implicated in this outbreak was a single brand of infant dried milk powder. Forty-six infants with symptoms had consumed the implicated product and 14 infections were associated with sibling or adult contacts of these cases. Two adults who had symptomless infection had consumed other products from the same factory where the implicated product was manufactured.

The outbreak strain was *Salmonella* Ealing (Table 2.8), which, following an intensive sampling exercise of products throughout the UK involving 4554 samples from 658 batches of product, was isolated from one opened packet of infant formula milk retrieved from the home of an infected infant. Subsequent to this, *S.* Ealing was found in four out of 267 sealed packs bearing the implicated batch code. The estimated level of contamination derived from this sampling exercise was calculated to be 1.6 organisms per 450 g of product (Rowe *et al.*, 1987).

A major investigation of the factory met with initial success as *S.* Ealing was recovered from scrapings taken from a waste powder silo where powder dust extracted from the vacuum system in the factory was collected. The organism was also isolated from several 25 kg bags of material from this silo. *Salmonella* was not found in samples of water or other environmental or product samples taken at this time in the factory.

The infant formula was produced from a blend of dried milk powder and other ingredients. The milk powder was produced on site. Raw milk was pasteurised and then concentrated through an evaporator. This concentrate was then spray dried and the powder filled into 25 kg bags for dispatch to other customers or for blending into the infant formula. An inspection of the records revealed no process deficiencies in the pasteurisation and evaporation stages and it was believed that these operations were under good control and not associated with this outbreak.

Table 2.8   Outbreak overview: infant formula milk

Product type:	Spray dried milk powder used for infant formula
Year:	1985
Country:	England
Levels:	1.6 organisms/450 g
Organism:	*Salmonella* Ealing
Cases:	76 (1 death)

Possible reasons
(i)   Use of water from a poorly covered storage tank accessible, to birds and external contamination, to test wet-down integrity of spray drier and leakage of water and contaminants into the insulation cavity between inner and outer skin of the spray drier
(ii)   Hole in inner skin of spray drier allowing milk powder into the moist insulation cavity, proliferation of contaminants and re-introduction of contaminants into the spray drier
(iii)   Consumption of the product by vulnerable groups

Control options*
(i)   Effective maintenance of the integrity of equipment
(ii)   Use of chlorinated, controlled sources of water for wet-down testing process
(iii)   Full HACCP assessment and consequent effective control of all potential external sources of contamination as well as internal sources

* Suggested controls are for guidance only. It is recommended that a full hazard analysis be carried out for every process and product to identify where controls should be implemented to minimise the hazard from *Salmonella*.

Although the silo used for collecting the waste powder from the vacuum system was found to be positive for *S.* Ealing, the source of contamination could not be determined readily. The vacuum system extracted powder from a variety of sites in the factory and samples were taken from all of the ducts. Powder scrapings taken from joints in all but one of the ducts were found to be contaminated with *S.* Ealing, although the primary source still remained a mystery.

Consequently, a full internal inspection of the plant was carried out which included the spray drier. The inspection revealed several pinpoint holes in the inner lining of the drier together with one large tear measuring 1 cm by 3 cm. The outer casing of the drier was removed at this point to reveal a large collection of discoloured milk powder together with staining of the insulation material between the inner and outer lining. *Salmonella* Ealing was detected in samples of powder and insulation material taken from this source. Other samples of insulation material taken from other sites of the drier did not yield the organism with the exception of a wet sample of insulation material taken from an area several metres below the hole.

Workers in the factory were all stool tested to determine carriage rates of the organism and no positive results were found in any of the current workers. However, a sample taken from a worker who had previously left the company was found to be positive for *S.* Ealing. The worker had been responsible for maintenance of the vacuum system, which involved opening and unblocking it as required. He reported suffering gastro-intestinal illness after one such occasion (Rowe *et al.*, 1987).

The origin of the *S.* Ealing in the factory was the subject of much speculation including possible entry in improperly pasteurised milk or via unrestricted access to personnel from the wet side of the factory to the milk powder plant. However, there was no evidence to substantiate such occurrences and, indeed, no deficiencies in these practices were noted. The most likely source of contamination was thought to be from the top of the spray drier, *Salmonella* being introduced during major clean downs. The top of the spray drier was formed from interlocking steel plates and it was possible for the water to seep between the joints, transferring microbial contaminants from the external environment into the insulation material between the inner and outer skin of the drier (Rowe *et al.*, 1987). It was also a reported practice for the factory to test the spray drier wet-down safety system using water stored in a partially covered tank in a roof-top building which was accessible to birds. *Salmonella* Ealing was not isolated from this water or bird droppings taken from this source, but it was still considered the most likely route by which the organism first gained access to the spray drier.

Water has been implicated as a possible source of contamination in out-breaks of salmonellosis involving other dried products. A large outbreak of salmonellosis in England and Scotland affecting 47 people in 1989 was attributed to autolysed yeast powder used in flavouring for savoury corn snacks (Joseph *et al.*, 1991). The yeast powder was produced using a roller drier, and the outbreak strain, *S.* Manchester, was found in batches of yeast powder, in the roller drying area and in the drainage system from the roller drying building. *Salmonella* Oranienburg, *S.* Schwarzengrund and *S.* Manchester were all isolated from drainage water taken from a local stream, which was reportedly used in the factory as a coolant (Joseph *et al.*, 1991). It is not clear whether this water was the original source of the contamination or whether it became contaminated from the factory.

Salmonellosis associated with milk powder has been reported on a number of occasions and clearly demonstrates the elevated risk associated with this material and products made from it. An international outbreak of infection due to *S.* Anatum was linked to infant formula milk in 1996-7 (Anon, 1997a). Cases of illness were recorded in England and France and *S.* Anatum was isolated from an unopened sachet of formula milk retrieved from the home of one of the affected infants. An outbreak of salmonellosis reported in Canada and the USA in 1993 affected three infants following the consumption of powdered infant formula milk contaminated with *S.* Tennessee (Louie *et al.*, 1993). This outbreak was unusual in that the isolated strain fermented lactose. Many laboratories may have overlooked this isolate as lack of lactose fermentation is often used to differentiate *Salmonella* from other members of the Enterobacteriaceae (Tables 1.1 and 1.2). The strain was isolated from stools of the infants, the powdered milk product, and from production equipment at the factory in Minnesota, USA where it had been produced.

The elevated microbiological safety risk associated with these products arises for three principle reasons. Firstly, the production process is complex, involving the processing of raw milk through pasteurisation and a drying system that is vulnerable to the introduction of environmental contaminants. Secondly, these products are intended for the most vulnerable group of individuals in the community, babies. Finally, the product is reconstituted and usually warmed, which, depending on the practices of the parent, could offer growth opportunities for any bacterial contaminants.

A key objective of the manufacturing process for milk powder production is clearly to prevent pathogens such as *Salmonella* being present in the finished product. The process starts with the delivery of milk. In the UK infant dried milk outbreak it is reported that milk was delivered raw and subjected to on-site pasteurisation. From time to time, raw milk will be contaminated with *Salmonella* and other enteric pathogens albeit, in most cases, at a low level and frequency. Strict procedures must therefore be in place to segregate raw milk from all subsequent stages of processing to avoid any chance of cross-contamination. Systems must also be in place to prevent staff employed on the raw milk side from transferring contamination from the raw side of the plant to the processed side. It has been suggested that spray drier plants should not receive and handle raw milk products but instead, should use only previously pasteurised materials in order to reduce the chance of pathogens entering the production facility (Rowe *et al.*, 1987). Nevertheless, pasteurisation temperatures should be employed that will destroy contaminants such as *Salmonella* in the milk; this is readily achieved using the European minimum standard of 71.7°C for 15 seconds (Anon, 1992b). Like any pasteurisation plant for milk, the process should be subject to extensive process control and monitoring. Divert valves should be in operation on the pasteuriser plant and the thermal process should be monitored using chart recorders. Simple verification checks on the milk pasteurisation process should include routine analysis of the pasteurised milk for phosphatase, an enzyme that indicates inadequate pasteurisation if detected.

After pasteurisation, milk is condensed through a series of chambers in an evaporator. The evaporator operates under vacuum to depress the boiling point and therefore retain optimum nutritional and organoleptic qualities of the milk. Care must be taken in the operation of the evaporator due to the potential for growth of surviving microorganisms during this process. Few vegetative bacteria would be expected to be present in the milk, but spore formers and thermoduric bacteria may remain and, as the latter stages of evaporation may fall within the growth range of these organisms, this needs to be carefully controlled. Particular attention must be paid to the cleaning of evaporator systems together with subsequent buffer tanks and pipework prior to the drier as poor hygiene can lead to a build up of microorganisms in these parts of the plant. These sites should be subject to routine sampling programmes to monitor cleaning efficacy.

After the evaporator, the concentrate is fed to the spray drier entering the drying chamber as a fine (atomised) stream of liquid where it encounters hot air which enters the chamber from direct gas or steam heaters. The liquid droplets formed by the atomiser and/or the interaction between the liquid stream and hot air, dry rapidly to form milk powder. Dried milk powder is collected at the bottom of the drying chamber or by an attached cyclone chamber, most units incorporating the latter to collect smaller powder particles that exit with the now warm, moist air.

Depending on the nature of the product, milk powder may be subject to further drying on a fluidised bed at the base of the drier or it may be re-wetted with steam and re-dried; this facilitates improved solubility in 'instant' product recipes. As the moisture level of the milk powder does not increase beyond 10–12% during these stages and, indeed, is less than 8% in the final product, there is little opportunity for microbial growth during these latter stages unless local pockets of moisture build up in any of the vessels or pipes.

Hot air entering the drier reaches an ingoing temperature between 180°C and 250°C but it is important to recognise this is not the temperature to which the milk concentrate is raised. The objective of spray drying is to remove moisture and not to cook the milk droplets, otherwise, unacceptable flavour changes occur. Therefore, whilst some reduction in viable microbial numbers can be achieved during the spray drying process, studies have shown that vegetative contaminants entering in the concentrate do survive in the finished powder (International Commission on Microbiological Specifications for Foods, 1980).

Microbial contaminants can gain access to the milk powder from a wide variety of sources, e.g. from air used to heat and cool the powder in the drier or cyclones, from the surfaces of production equipment such as the drier walls, fluidised beds and silos or from the general environment and personnel. Air used directly in powder production facilities must be filtered using systems capable of removing microbial contaminants. The efficiency of these filters is critical to the safety of the product and so should be effectively maintained and controlled. With consistently effective control of air and milk quality entering the drier, the powder plant is then most vulnerable to introduction of contamination from the processing environment and people. Procedures must be in place to ensure cross-contamination does not occur from the raw side of the factory and strict personnel control should be in operation in milk powder production plants. The drying plant should be completely physically segregated from other parts of the plant, with product being fed in through enclosed pipework. The environment must be under positive air pressure and any entrances and exits should be designed as a double-door system that helps to maintain the positive pressure from the powder plant to any other area during staff entry and exit. Personnel (operatives, engineers and visitors) should completely change all outer clothing in a high-risk changing facility and wear overalls, hair covering and footwear dedicated for use only in the powder plant. Clearly, these procedures should be supported by a stringent infectious disease policy for new starters, staff suffering intestinal illness or in contact with those suffering illness and also, staff returning from holiday abroad.

The source of *S.* Ealing in this outbreak appears to have been the equipment in contact with the product, namely the spray drier, although it is not known how the organism got there. All equipment downstream of milk pasteurisation must be considered to represent a hazard to the finished milk powder product. As such, it should be subject to regular, effective cleaning and disinfection programmes supported by routine inspection and planned preventative maintenance. This includes the pasteuriser, evaporator, buffer tanks, atomisers, spray drier, fluidised bed, silos and bagging units and all associated transfer pipes. Clearly, this is no insignificant undertaking but the controls necessary to ensure safety of this most vulnerable product can be effectively identified and controlled using a proper and thorough HACCP-based approach. Each piece of equipment should be assessed in relation to the hazards presented to it and by it and the required controls must be implemented.

The holes in the spray drier and the associated cross-contamination between the insulating material and the contents of the drier are perfect examples of the need to conduct regular assessments of the integrity of all equipment. This should be aimed at preventing access of contaminants to the equipment in the first place, preventing build up of contaminants if they were to occur and, critically, ensuring such contaminants cannot gain access to the product.

The degree of control in place in the processing of milk powder usually precludes opportunities for *Salmonella* to enter the product through the milk itself and access is usually gained via people or the environment. It is therefore, common practice for the results of microbiological examination of the equipment and environment to be used as a monitor of the potential for contamination of the final product. In addition to a very large number of finished product powder samples taken for examination for *Salmonella* and other indicator microorganisms such as *E. coli* and Enterobacteriaceae, a much larger number of environmental samples are usually taken from the surrounding processing environment. This includes floor sweepings in the powder plant including vacuum dust, samples of air pre-filters and filters, water samples and swab samples. It should also include sources external to the factory, particularly roofs and roof spaces.

Extensive product sampling is usually carried out in milk powder production with samples being examined specifically for the presence of *Salmonella*. The level of testing required clearly needs to be established for each individual circumstance. However, these days, in-line sampling devices are used that continuously remove small quantities of product flowing past the sampler; this method considerably increases the chances of detecting any contaminating microorganisms. It should, however, always be remembered that any sampling exercise is just that, a sample, and it cannot give absolute confidence in the safety of the product. This can only be achieved through the identification and implementation of effective controls from the receipt of the raw material to the despatch of the final product.

Once present in the dry milk powder, *Salmonella* is known to survive for periods in excess of a year, and often well beyond the shelf life of the product, thus presenting a prolonged hazard.

It should also be remembered that the final product, infant formula milk, will be reconstituted and warmed by the parent. Instructions present on the packaging of such products must preclude the possibility that the warmed milk product will be held for too long a time before use. Although this would be undesirable with regard to the potential growth of pathogens such as *Salmonella*, clearly such organisms should not be present in the first place. However, it should be remembered that milk powder will contain low levels of surviving spore-forming pathogens such as *Bacillus* species which could also grow and present a hazard if the reconstituted powder is held for too long a time at a warm temperature.

The vast quantities of milk powder produced and consumed each year without apparent ill effect are testimony to the ability of processors to apply high standards of process control and hygiene. But the safety of these products must be earned through the continued investment in equipment, people and procedures and not through historical precedence.

PASTEURISED MILK CHEESE: ENGLAND

In December 1996, an increase in the number of isolations of an uncommon *Salmonella* serotype was noted in the Northern and York-shire health region of England. A case control study of 15 cases indicated a strong association with the consumption of a single food product (Anon, 1996a).

The outbreak strain was isolated from 84 cases who suffered illness between August 1996 and February 1997 and, of 27 cases interviewed, nine recalled definitely consuming the implicated product (Anon, 1997b).

The product was a mild, coloured, Cheddar cheese, manufactured from pasteurised milk and sold in bulk packs ($4 \times 5\,$kg or $8 \times 2.5\,$kg) for subsequent slicing on delicatessen counters or for individual pre-packed units ($220\,$g, $227\,$g or $250\,$g). The product was manufactured in the south of England and sold under a variety of names, although the implicated batch was principally distributed to outlets in the north of England. The UK Department of Health issued food hazard warnings on 16 and 18 December (Anon, 1996b) and the implicated product was recalled from sale.

The outbreak strain was identified as *Salmonella* Goldcoast (Table 2.9). This strain was isolated from cheese manufactured in August and September 1996 and from a dairy herd supplying milk to the cheese manufacturer.

Cheese, of both pasteurised and unpasteurised varieties, has been implicated in a large number of outbreaks of salmonellosis. In many cases it is the use of unpasteurised milk, either intentionally or unintentionally that appears to have been the most significant factor in the cause of outbreaks. In 1984 a large outbreak of salmonellosis caused by *S.* Typhimurium PT10, affecting more than 1500 people, was reported in Canada (D'Aoust *et al.*, 1985) due to the consumption of Canadian cheddar cheese made from pasteurised or thermised (heat treated but not fully pasteurised) milk (D'Aoust, 1989). In 1989, imported Irish soft cheese made from unpas-teurised milk was responsible for salmonellosis affecting 42 people in England and Wales (Maguire *et al.*, 1992). The outbreak strain was *S.* Dublin and this was isolated from samples of cheese taken from the manufacturer's premises. The manufacturer subsequently modified the process to pasteurise milk for the production of the cheese. Cody *et al.* (1999) reported two outbreaks of *S.* Typhimurium DT104 affecting 31 and 79 people, respectively, following the consumption of unpasteurised milk

Table 2.9 Outbreak overview: pasteurised cheese

Product type:	Mild Cheddar cheese made from pasteurised milk
Year:	1996/7
Country:	England
Levels:	Not reported
Organism:	*Salmonella* Goldcoast
Cases:	84 (no deaths)

Possible reasons
(i)   Inadequate pasteurisation of milk
(ii) Inadequate response to process control and quality control results
(iii) Survival of the organism in cheese given a short maturation time

Control options*
(i)   Training of personnel in proper operation of pasteuriser
(ii) Daily review of process records such as thermographs by a designated responsible person and clear action procedures in the event of failures in process control and quality control test results, e.g. pasteurisation temperature/time and phosphatase test

* Suggested controls are for guidance only. It is recommended that a full hazard analysis be carried out for every process and product to identify where controls should be implemented to minimise the hazard from *Salmonella*.

Mexican-style cheese in northern California. The outbreak strains were multi-drug resistant, i.e. resistant to ampicillin, chloramphenicol, streptomycin, sulfonamides and tetracycline.

Altekruse *et al.* (1998) reviewed cheese-associated outbreaks of illness in the USA between 1973 and 1992 and reported that three outbreaks had been caused by *Salmonella*. In 1981, *S.* Typhimurium was implicated as the causative organism in an outbreak associated with mozzarella cheese. The outbreak was attributed to improper pasteurisation of milk and it resulted in 321 cases of illness in New York. One person was hospitalised and two died. In an outbreak identified in 1986, in Colorado, USA, illness was associated with the consumption of uncured cheddar cheese made from improperly pasteurised milk; 66 people were hospitalised out of a total of 339 cases. In 1989, a large multi-state outbreak affecting 164 people was attributed to mozzarella again, although on this occasion the cause was believed to be post-pasteurisation contamination, possibly from inadequate maintenance of shredding equipment (Altekruse *et al.*, 1998).

More recently, an outbreak of *S.* Typhimurium illness affecting 113 people was reported in France due to the consumption of a soft ripened cheese made from raw milk (De Valk *et al.*, 2000). *Salmonella* Typhimurium was isolated from products retrieved from patients' homes and the home of one individual not suffering illness.

Cheese is manufactured by the fermentation of milk using starter cultures. Milk, which may be pasteurised or unpasteurised, is warmed in a fermentation vessel to *c.* 30°C then lactic acid bacteria are added as starter cultures. These organisms multiply rapidly and ferment the lactose to produce acid with a concomitant decrease in pH. Rennet is added to the fermentation mixture and after about an hour this causes coagulation of the milk proteins to form a soft curd and liquid whey. The curd is usually cut in the vat and acid production continues by the activity of the starter cultures. In cheddar cheese production, the curd is cut into small blocks in the vat; these float in the whey and are stirred. At the same time, the temperature of the curd is increased to *c.* 40°C in a 'scalding' process that allows acidity and textural changes to develop. The whey is then drained and the curd pieces coalesce. The curd is then milled (broken down) into small pieces and salt is added. The salted pieces are deposited into cheese moulds and pressed to expel further moisture. This is done at ambient temperature over several hours. The cheeses are then removed from the moulds and placed into maturation rooms at 8–12°C and left to age for periods of up to a year or more. Flavour is significantly affected by the action of bacterial and milk enzymes and therefore the longer the maturation period the stronger the cheese flavour. Mild cheese is matured for short periods (2–3 months), whereas the full-flavour, extra mature varieties can be matured for periods beyond 12 months.

The reasons for the outbreak associated with mild cheddar cheese in England became quite clear in the subsequent investigation of the procedures in operation at the dairy. This outbreak was caused by an apparent failure in pasteurisation of the milk used for the cheese making process. The failure was detected by the monitoring systems in place, i.e. process records and verification checks, but the information was not apparently used to make the necessary corrective actions (Anon, 1997b).

Raw milk will be contaminated occasionally with enteric pathogens including *Salmonella* as dairy cattle can frequently shed enteric pathogens including *Salmonella* and minimising this organism's passage into milk is essential. A survey of faecal shedding of *Salmonella* by dairy cattle (1 g faecal samples) in the USA found 5.4% (198/3640) of milk cows, 18.1% (121/668) of milk cows due to be culled within 7 days and 14.9% (341/2287) of culled dairy cows to be shedding *Salmonella* (Wells *et al.*, 2001). It was estimated that *Salmonella* shedding in milk cows occurred at 21.1%

of dairies and 66% of cull dairy markets. The incidence and levels of enteric pathogens in milk can be reduced by effective dairy parlour hygiene practices but they cannot be totally excluded. Disinfection of cow teats and effective cleaning of the milking equipment will reduce the potential for pathogens to enter the milk supply, but this cannot guarantee absence of contaminants arising from faecal residues or asymptomatic infection of the udder.

In the manufacture of cheese made from unpasteurised milk, milking hygiene is probably the most critical stage contributing to the overall safety of the product. As such, sources of milk for the production of unpasteurised milk cheese should be under regular review in relation to adherence to high standards of hygiene. This should be supported by regular microbiological testing of the ex-farm milk for the presence of pathogens and indicators of faecal contamination such as *E. coli.*

The implicated product in this outbreak was supposed to be made from pasteurised milk. Therefore, whilst raw milk quality would still have been important, the critical control in relation to safety would have been the pasteurisation process. Pasteurisation, if carried out properly, is known to be capable of destroying *Salmonella,* if present in the raw milk (D'Aoust *et al.*, 1987). In the UK, pasteurisation of milk is achieved by a process equivalent to 71.7°C for a minimum of 15 seconds. It is believed that the manufacturer of the cheese implicated in this outbreak was working to this standard. Modern pasteurisation equipment is fitted with monitoring equipment capable of continuously assessing the pasteurisation temperature and diverting milk away from forward flow and into a re-circulation loop or a divert tank should the temperature of the milk not reach the desired level. The temperature at which divert occurs is usually set slightly higher than the minimum of 71.7°C to give a margin of safety. This is recorded on a continuous chart recorder or thermograph that is usually checked by quality control personnel during production or at the end of a day to verify that appropriate targets have been achieved. Because of the critical nature of pasteurisation, a further process efficacy verification check is carried out; this is by the use of the phosphatase test. This test detects the presence of the enzyme, alkaline phosphatase, that is inactivated by the thermal process used in normal milk pasteurisation. The detection of phosphatase in recently pasteurised milk, therefore, gives an indication of an inadequate heat process.

A full account of the causes of this outbreak has not been published. However, such reports as exist indicate that the manufacturer had monitoring systems in place that, if properly reviewed and acted upon, should have allowed an inadequate pasteurisation process to be detected. Thermograph recording of the pasteuriser temperature was in place together with phosphatase checks of the milk. In both these cases, results indicated that pasteurisation had not been effective on a number of occasions but it is not clear why the results were apparently not actioned thus preventing the outbreak. However, it is incumbent on those operating in the food industry to ensure that such checks are supported by appropriate management review of all results with clear action plans documented and enacted in the event of non-compliance.

A common reason why pasteurised milk becomes contaminated with unpasteurised milk is pasteuriser failure coupled with an ineffective flow diversion system to prevent improperly pasteurised milk continuing in forward flow. This may occur due to failure of the flow diversion valve or, more commonly, due to manual override of flow diversion. Many older pasteurisers have a manual override system, principally to allow forward flow during clean in place (CIP) cycles. Manual override of the pasteuriser during milk pasteurisation must never occur during the milk pasteurisation operation. In small factories, operatives often have multiple duties but it is essential that the operating pasteuriser remains under constant supervision by properly trained staff because flow diversion may require operative intervention, e.g. switching the raw milk pump off to prevent overflowing of the diversion tank when milk is sent into it.

Contamination of pasteurised milk may also occur through improperly maintained pasteuriser plates, with raw milk entering the pasteurised milk stream through cracks/pinholes in the pasteuriser plates.

In addition, in small farmhouse operations, it is also possible for cross-contamination to occur via the operatives themselves. As they often have several duties, it is not uncommon for operatives to handle raw milk pipes, tanks and other connections and then also set up and operate the pasteuriser or even handle the starter culture and inoculate the warmed milk in the fermentation tank. Significant opportunities exist for cross-contamination via these routes and systems of control that involve changing of footwear, clothing and washing of hands must be in place if operatives handle both raw milk and associated equipment and then the pasteurised product and its associated equipment.

Whatever the fundamental cause of this outbreak, it is essential that systems are in place to prevent raw milk from contaminating pasteurised milk through properly used and effective flow diversion systems, effective equipment maintenance programmes and prevention of physical cross-contamination via personnel practices.

It is vitally important that all persons involved in critical aspects of food process control are thoroughly trained and understand the process, reasons for and methods of control. Also, they must be fully aware of their responsibilities and completely understand the implications of any deviation from proper practice.

There are several other stages during the manufacturing process where cheese can become contaminated but the most critical point for pasteurised milk cheese is the milk pasteurisation stage. In addition to control of effective pasteurisation times and temperature, pasteurisers should be routinely tested for leakage between the plates. Post-process contamination due to poor segregation of raw and pasteurised milk or personnel with access to both sides of the factory are also important elements to consider in the control of these products. Environmental hygiene and effective cleaning and sanitisation of all equipment including the cheese vats, moulds and maturation rooms is also critical to avoiding contamination entering at these points of the process.

It is interesting to note that this outbreak was caused by a hard cheese, as it is well known that enteric pathogens do reduce in number during the maturation stages of hard cheese products. However, this cheese was a mild variety and it is probable that the maturation time was insufficient to allow a significant reduction in the level of any *Salmonella* present. Indeed, this would have been compounded by the fact that levels of *Salmonella* in cheese capable of causing illness are thought to be very low due to the protection afforded to the organism by the high fat level during passage through the stomach. Following the large outbreak in Canada caused by cheddar cheese, D'Aoust *et al*. (1985) reported that levels in the cheese ranged from 0.36 to 9.3 *S*. Typhimurium PT10 per 100 g cheese. In a study of the contaminated cheese batches, initial levels of 0.3 to 2.3 *S*. Typhimurium per 100 g cheese remained detectable for up to 8 months at 5°C (0.7 *S*. Typhimurium per 100 g cheese) (Table 2.10).

Park *et al*. (1970) reported on studies on the survival of *S*. Typhimurium in a low-acid cheddar cheese. Cheese was manufactured from pasteurised milk inoculated with *c*. 10^2 *S*. Typhimurium/ml. The milk was fermented using added starter culture at 31.2°C prior to adding rennet to coagulate the proteins. Four hours after the addition of the starter culture, the whey was drained and the curd was salted over the following 1.5 hours. The cheese was filled into moulds and entered the press approximately 6

Table 2.10 Survival of *S.* Typhimurium PT10 in naturally contaminated Cheddar cheese, adapted from D'Aoust *et al.* (1985)

Vat no.	pH	Age of cheese (months) when sampled*								
		2	3	4	5	6	7	8	9	10
89	5.22–5.4	NT	NT	NT	NT	0.91†	—	—	NT	NT
299	5.18	NT	4.3	0.9	0.3	4.3	1.5	0.7	—	—
314	5.18	NT	0.9	4.3	—	0.9	0.7	—	—	NT
475	5.09	0.9	—	0.3	—	—	NT	NT	NT	NT
495	5.11–5.14	2.3	0.3	0.3	0.3	—	—	NT	NT	NT
530	5.11–5.18	0.3	—	—	NT	NT	NT	NT	NT	NT

* Cheeses of different ages were taken from the implicated manufacturer and distributor and after initial sampling were stored at 5°C and tested for the organism. The first result for *Salmonella* in each column corresponds to the initial age of the cheese when taken.
† Most probable number (MPN) of *Salmonella*/100 g cheese.
— = negative (not detected).
NT = not tested.

hours after the addition of starter culture; the cheese was then pressed for about 18 hours at room temperature. After this time the product was stored at 7°C or 13°C for up to 10 months.

The pH of the cheese after pressing in five trials carried out ranged from 5.65 to 5.82 and the moisture content 42.5% to 44.8%. In both cases these are rather higher pH and moisture content levels than would normally be expected at this stage during typical cheddar cheese manufacture. Levels of *S.* Typhimurium inoculated into the milk between 140 and 600/ml increased during the early stages of the process to 5.6×10^3/g–2.9×10^4/g in curd after draining and to 1.5×10^4/g– 1.1×10^6/g in cheese after pressing. Although some of this increase may be attributable to concentration of the organism due to whey removal, this still indicates significant growth over this period, most likely due to the slow pH decrease and acidity development in the cheese as a result of slow starter activity. In cheese stored at 13°C, over the following 4 weeks, the pH of all cheeses decreased gradually to *c.* 4.9–5.1 and, with the exception of one cheese where the pH increased to *c.* 5.5 after seven months storage, all others remained at a fairly constant, low pH. The levels of *S.* Typhimurium declined by *c.* 1–2 log units in the first 4 weeks of storage at 13°C and, with the exception of the cheese where the pH began to elevate, levels after two months had dropped to log 2 cfu/g or less. By four months, the organism was not detectable in all cheeses, with the exception of the cheese with elevated pH where the organism remained at *c.* log 1 cfu/g up to month 7. At 7°C, the time to achieve low pH (<5.4) was significantly extended, i.e. from four weeks to four months, and only two cheeses reached a pH below 5.2. Survival of *S.* Typhimurium was also very erratic with levels decreasing by *c.* 1-2 log units within four weeks. Levels generally became undetectable after 6-7 months but, in one cheese, the organism was still present after 10 months. This work clearly demonstrates the importance of pH and acidity brought about by active fermentation, and time on the survival of *Salmonella* in cheese.

The continued outbreaks of salmonellosis implicating cheese demonstrates the vulnerability of these products to contamination by *Salmonella*. However, it should be possible to prevent such outbreaks, especially in pasteurised milk varieties, through the application of appropriate hygiene controls in the raw milk supply, pasteurisation process control, segregation of raw milk from the maturing and finished product and effective hygiene management of the processing environment and equipment.

CATERING AND THE HOME

A large number of outbreaks and incidents of salmonellosis occur as a result of food incorrectly prepared in the home or in semi-professional catering or retail operations. A selection of these outbreaks and factors contributing to them are detailed collectively in the following text.

(i) Home-made ice cream made with raw eggs

In October 1996, 37 people, seven adults and 30 children attended a birthday party in north London, England. Within 24 hours, 30 people developed symptoms of gastroenteritis; 27 had diarrhoea, 25 fever, 21 vomiting and nine abdominal cramps; illness lasted between two and three days (Dodhia *et al.*, 1998). The children were mainly four to five years of age and one child was admitted to hospital due to dehydration.

The implicated food vehicle was ice cream (Table 2.11), which was the only home-made food item consumed at the party. It had been prepared on the day prior to the party using a recipe of raw shell eggs, sugar and a whip topping and the ice cream was stored frozen until consumption. No information was available concerning the source or storage conditions of the eggs. *Salmonella* Enteritidis PT6 was isolated from 24 out of 25 stool specimens submitted for analysis, including one from the hostess, who was asymptomatic. None of the ice cream was available for analysis and

Table 2.11   Outbreak overview: home-made ice cream

Product type:	Home-made ice cream made with raw eggs	
Year:	1996	
Country:	England	
Levels:	Not reported	
Organism:	*Salmonella* Enteritidis PT6	
Cases:	37 (no deaths)	

Possible reasons
(i)   Use of contaminated raw eggs in ice cream
(ii) Survival of *Salmonella* in frozen product
(iii) Consumption by vulnerable groups (children) at a party

Control options*
(i)   Use of pasteurised egg

* Suggested controls are for guidance only. It is recommended that a full hazard analysis be carried out for every process and product to identify where controls should be implemented to minimise the hazard from *Salmonella*.

the level of contamination could not be ascertained but the use of raw eggs in this recipe is most likely to have been the cause of the outbreak.

A family outbreak of salmonellosis occurred in England (Morgan *et al.*, 1994) due to the consumption of home-made ice cream. The ice cream had been made from raw eggs and analysis of the ice cream found *S.* Enteritidis at levels of *c.* 10^5/g.

A further outbreak was reported in Florida, USA in 1993 (Buckner *et al.*, 1994) where 12 out of 14 people suffered salmonellosis 7–21 hours after eating food at a 'cookout' at a psychiatric treatment hospital in Jacksonville. Five of those affected were children and seven were adults. Eleven of the 12 ill persons had consumed home-made ice cream served at the event. *Salmonella* Enteritidis PT13a was isolated from the three stool samples obtained from affected individuals. The same strain was also isolated from ice cream which had been prepared using raw eggs, three hours before the meal. The person who prepared the ice cream had no previous history of salmonellosis but became ill 13 hours after eating some of the product.

It seems likely that use of pasteurised eggs in all these ice cream recipes would have avoided these outbreaks.

Ice cream was also the vehicle responsible for one of the largest salmonellosis outbreaks in the USA caused by a commercially manufactured product (Table 2.12). In 1994, an estimated 224 000 people nation-wide suffered illness caused by *S.* Enteritidis (Hennessy *et al.*, 1996). The organism was isolated from samples of three production lots of ice cream that were subjected to extensive microbiological examination to determine the numbers present. It was estimated that the number of *S.* Enteritidis cells per serving (65 g) was 25 and that the infectious dose was no more than 28 cells (Vought and Tatini, 1998).

It was believed that the outbreak was caused by the contamination of pasteurised ice cream mix during transport to the factory. The tanker trailers used to transport the ice cream premix had previously carried unpasteurised liquid egg and, because the premix was not subject to re-pasteurisation during the ice cream manufacturing process, any contaminants remained in the final product. Table 2.12 indicates the basic control measures that, if implemented, are likely to have been effective in preventing this outbreak.

Table 2.12 Outbreak overview: commercial ice cream

Product type:	Commercially produced ice cream
Year:	1994
Country:	USA
Levels:	25 cells per 65 g product
Organism:	*Salmonella* Enteritidis
Cases:	224 000 estimated (no deaths)

Possible reasons
(i)   Transportation of pasteurised ice cream premix in trailer previously used for raw egg
(ii) Inadequate cleaning/sanitisation of trailer between loads
(iii) Use of material in product that was not subjected to a further pathogen destruction stage

Control options*
(i)   Effective cleaning and sanitisation of trailer
(ii) Use of dedicated trailer for pasteurised product
(iii) Packing of pasteurised premix in units capable of being sanitised after transportation and into production or, with outer packaging capable of being removed hygienically prior to intake
(iv) Re-pasteurisation of premix immediately prior to ice cream production

*Suggested controls are for guidance only. It is recommended that a full hazard analysis be carried out for every process and product to identify where controls should be implemented to minimise the hazard from *Salmonella*.

(ii) Take-away kebabs with yoghurt dressing

An outbreak of illness caused by *S.* Typhimurium DT170 affecting up to 52 people was reported in South Wales (Evans *et al.*, 1999). The implicated food was kebabs (either doner or shish) with yoghurt-based dressing (Table 2.13). Kebabs are manufactured from raw lamb, either chopped into small cubes and grilled or barbecued on metal or wooden skewers (shish kebab), or minced and formed into large cylindrical meat joints (*c.* 18–24 inches (*c.* 45–60 cm) long by 8–12 inches (*c.* 20–30 cm) in diameter) and then grilled on a rotisserie with slices being cut off to order (doner kebab).

Two retail outlets were implicated in this outbreak, both of which were supplied by a common wholesaler. Practices in both outlets were investigated. No major deficiencies were noted at one of the retail outlets, but potential for cross-contamination from raw chicken and poor temperature control was noted at the second retail outlet. The outbreak was considered to have been most likely caused by major deficiencies revealed at

Table 2.13   Outbreak overview: kebabs

Product type:	Take-away kebab with yoghurt dressing
Year:	1995
Country:	Wales
Levels:	Not reported
Organism:	*Salmonella* Typhimurium DT170
Cases:	52 (no deaths)

Possible reasons
(i)   Storage in the same refrigerator of raw lamb carcasses above containers of yoghurt
(ii)   Blood drip onto/into poorly sealed yoghurt cartons
(iii)   Use of contaminated yoghurt for dressing cooked meat kebabs
(iv)   Undercooking of kebab

Control options*
(i)   Effective segregation of raw meat from ready-to-eat foods
(ii)   Secure seal integrity of ready-to-eat product containers
(iii)   Avoidance of the use of evidently contaminated ready-to-eat product, i.e. blood drip
(iv)   Thorough cooking of kebab meat

* Suggested controls are for guidance only. It is recommended that a full hazard analysis be carried out for every process and product to identify where controls should be implemented to minimise the hazard from *Salmonella*.

the premises of the wholesaler that were described as dirty and structurally unsuitable (Evans *et al.*, 1999). Lamb carcasses were hung in a refrigerated room at the back of the wholesale premises, the front of which was a retail shop. Yoghurt was stored in uncovered boxes in the same refrigerator, directly underneath the lamb carcasses. Several yoghurt cartons, which had loose-fitting lids, were observed to be contaminated by blood, presumably dripping from the raw lamb carcasses. *Salmonella* Typhimurium DT170 was isolated from minced lamb used as kebab meat and from food debris taken from a number of areas at the wholesale supplier's premises (ice scrapings from a freezer, scrapings from a table top and from a storage room floor).

It is difficult to believe that this outbreak was not caused, or contributed to, by cross-contamination of the yoghurt from blood dripping onto and probably into it from the raw lamb carcasses whilst being stored together in the refrigerator on the premises of the wholesaler, although undercooking of contaminated kebab meat could also have been a potential factor.

Implementation of basic food safety practices including adequate separation of raw and ready-to-eat products in clean store rooms and refrigerators may have prevented this outbreak.

(iii) Home-cooked cockles supplied to a fish and chip shop

In the UK in July 1997, a cluster of infections caused by *S.* Typhimurium PT19 was noted in eight people who were all subsequently found to have consumed cockles (Greenwood *et al.*, 1998) (Table 2.14). The cockles were purchased by the affected individuals from a local fish and chip shop or received them directly from the supplier. The supplier was a local resident who had collected cockles from the harbour shore and then stored them in a refrigerator overnight. The cockles were boiled in a domestic saucepan, after which they were held in cold water while removing the flesh from the shells. It is reported that the resident had collected surplus cockles for her domestic use and so had placed the remaining cockle flesh into small containers, transported them in a cool box and offered them for sale by the fish and chip shop.

Table 2.14 Outbreak overview: cooked cockles

Product type:	Home-cooked cockles sold to a retail outlet
Year:	1997
Country:	England
Levels:	Not reported
Organism:	*Salmonella* Typhimurium PT19
Cases:	8 (no deaths)

Possible reasons
(i)   Harvesting of cockles from contaminated and/or uncontrolled water sources
(ii) Undercooking of cockles or cross-contamination of cockles post cooking
(iii) Inadequate refrigeration of cooked cockles

Control options*
(i)   Harvesting shellfish from controlled water sources
(ii) Effective cooking and prevention of cross-contamination through segregation of raw and cooked products, including the use of dedicated utensils for cooked cockles and effective cleaning/sanitisation of hands before handling cooked cockles
(iii) Effective temperature control of cooked product ($<8°C$)

*Suggested controls are for guidance only. It is recommended that a full hazard analysis be carried out for every process and product to identify where controls should be implemented to minimise the hazard from *Salmonella*.

The supplier had cooked cockles for domestic consumption for many years but could not specify the time and temperature of the cook, presumably not having any means to monitor the product temperature.

Symptoms occurred within 24 hours of consumption of the cockles although most of those affected were reportedly ill within 12 hours (Greenwood *et al.*, 1998). Individuals who ate whole portions of the cockles suffered severe diarrhoea and a single person was admitted to hospital. Other symptoms included vomiting, headache, abdominal cramps and fever, and the shop owner's daughter, who had only tasted but not consumed the product, suffered mild illness. No samples of the cockles were available for analysis but, given the nature of the outbreak, it is probable that high levels of *Salmonella* were present. Cockles supplied to the fish and chip shop were reported to be stored in a refrigerated glass cabinet or at ambient temperature during trading hours.

The contamination source was most likely to have been the cockles themselves which were harvested from the shoreline of a harbour. Cockles are filter-feeding molluscs capable of concentrating bacteria from their environment. Enteric pathogens including *Salmonella* arising from sewage outflows or even ships emptying tanks will also be filtered and concentrated by cockles. The cockles were either undercooked or, they could have been cross-contaminated after boiling during removal of the flesh via contaminated utensils and/or from handling raw and then cooked cockles without proper attention to hand-washing. The contaminants may then have increased in number during retail display under ambient conditions prior to their consumption. Table 2.14 indicates some simple control measures that should be effective in preventing outbreaks of enteric illness from these types of food products.

(iv) Cooked beef from a restaurant/caterer

Fifty-two *S.* Thompson infections detected in an outbreak between September and October 1996 implicated a restaurant in Sioux Falls, South Dakota, USA (Shapiro *et al.*, 1999). Illness was associated with the consumption of roast beef (Table 2.15) that had been prepared by a single restaurant and served at the same restaurant or supplied on platters to three local lunchtime functions. *Salmonella* Thompson infection was noted in four employees and five patrons of the restaurant together with 43 persons who were served the food at the three separate functions for which the food was provided by the restaurant. Of the four restaurant employees found to be infected with the organism, one was a chef who reported suffering gastrointestinal illness in early October, a second was a

Salmonella

Table 2.15 Outbreak overview: cooked beef

Product type:	Restaurant cooked beef
Year:	1996
Country:	USA
Levels:	4.6×10^3/g
Organism:	*Salmonella* Thompson
Cases:	52 (no deaths)

Possible reasons
(i) Undercooking of contaminated beef
(ii) Inadequate temperature control after cooking
(iii) Beef served without re-cooking

Control options*
(i) Source raw beef from suppliers operating high standards of hygiene
(ii) Effective cooking to achieve a minimum 70°C held for two minutes (UK standard)
(iii) Effective temperature control of cooked product (< 8°C)

*Suggested controls are for guidance only. It is recommended that a full hazard analysis be carried out for every process and product to identify where controls should be implemented to minimise the hazard from *Salmonella*.

delivery person who suffered illness in mid-October and the other two were another chef and a dishwasher/cook who suffered no reported symptoms.

Samples of leftover cooked beef and ham which were stored in separate plastic bags in a refrigerator following one of the luncheons were found to be contaminated with *S.* Thompson. Levels in the ham, determined by the most probable number technique (MPN), were 23 cfu/g and in the beef, 4600 cfu/g. The beef was reported to have a pink colour and it was speculated that it may have been undercooked (Shapiro *et al.*, 1999).

The roast beef was believed to be the cause of this outbreak with other foods probably being contaminated from this primary source. Two beef joints were routinely cooked by the restaurant for Sunday 'brunch'. The joints were taken out of refrigeration, after having been aged at 40°F (4.4°C) for 2–3 weeks, and then placed in an oven at the end of trading on Saturday night. On Sunday morning at 7.00 am the internal temperature of the roast was checked (110°F/43.3°C), then, over the next three hours, the temperature was slowly raised so that by 10.00 am it had reached 125–130°F (51.7–54.4°C). At this point the first slices were taken for serving in the restaurant. It should be noted that the overnight 'storage' of the beef at temperatures allowing active multiplication of *Salmonella* could have resulted in high numbers being present before the temperature was raised. After the 'brunch' period, the remaining meat portions were laid out to cool and then wrapped in cellophane and placed in a refrigerator reportedly operating at 50°F (10°C). This meat was then used for catering events and cold meals during the subsequent week. Meat was prepared for the three luncheons using the cooled beef, ham and turkey, all of which were sliced on the same slicer. Products were delivered to the functions at 10.15 am, 11.30 am and 11.40 am on the same day, and were believed to have been stored at room temperature for several hours before consumption (Shapiro *et al.*, 1999).

It was reported that this outbreak, attributable to undercooking of beef followed by poor post-cooking temperature control, could have been avoided if the restaurant was complying with the US Food and Drug Administration (FDA) Food Code. This indicates that the roast beef should have been cooked to an internal temperature of 130°F (54.4°C) maintained for two hours or to 145°F (62.8°C) and maintained for three minutes to achieve a 5 log reduction in *Salmonella*. In addition, a refrigeration temperature of 41°F (5°C) should have been employed to inhibit the multiplication of the organism, if present, during storage (Shapiro *et al.*, 1999).

The different outbreaks described here provide examples of many of the factors which, singly or combined, are often implicated in outbreaks of salmonellosis, i.e. the use of raw eggs in uncooked dishes, undercooking of raw meat or poultry, cross-contamination in the kitchen and some form of storage temperature abuse.

Outbreaks like these occur frequently often because there is a lack of understanding, by those preparing the food, of the hazards presented by certain food ingredients and/or a lack of implementation, through ignorance or non-compliance, of requisite procedures for their control.

The use of potentially contaminated ingredients such as raw shell eggs in the preparation of uncooked foods such as ice cream or dairy-based desserts is commonplace. Eggs are a well-known hazard in relation to *Salmonella* and the organism may be present both on the outer surface and in the contents of the egg. The incidence of *Salmonella* in eggs does vary but in the UK has been reported to range from one in every 650 eggs (de Louvois, 1993) to one in 2900 (de Louvois, 1994). Refrigeration of eggs after purchase can significantly reduce the colonisation and subsequent growth of *Salmonella* in the egg contents.

Whilst the use of raw egg and the consumption of raw egg dishes should remain a matter for individual choice, it is incumbent on those preparing and serving meals to others to inform them of the use of such materials so that they too can make an informed choice about what they eat. In the UK, government advice is to avoid the consumption of raw eggs and uncooked foods containing raw eggs (Anon, 1998e). This is particularly the case for vulnerable groups such as pregnant women, babies, the elderly, those who are ill and the immuno-compromised.

Undercooking of raw meat products is another common contributory factor in foodborne outbreaks and incidents of salmonellosis originating in restaurant, institutional and consumer kitchens. Sadly, this is also found as a contributory factor in some manufacturing industry associated outbreaks of salmonellosis.

Clearly, it is difficult to properly control the cooking process for raw meat and poultry in the home. However, public education and reliable information on how to cook raw meats and poultry thoroughly can help raise awareness of the importance of effective cooking and, hence, reduce the number of incidents of foodborne illness associated with undercooking of meats. Using communication vehicles such as the labels on the meat packaging itself or leaflets and advertising campaigns are effective ways of disseminating these important messages. These days, some consumers, particularly in the USA, use meat probes during home-cooking to check that the temperature of the meat being cooked has reached the correct level for ensuring the destruction of pathogens and therefore, the safety of the meat. This is not common practice in the UK. However, large restaurants, caterers and small manufacturers really have no excuse for not ensuring that appropriate temperatures are achieved during cooking, and meat temperature probes or other indicators of cooking efficacy should be used routinely. Cooked meat manufacturers should also ensure that appropriate cooking validation studies are carried out to ensure that the process routinely applied for cooking is capable of delivering the correct cook under worst-case conditions including consideration of ingoing meat piece size, temperature and position of meat products in the oven (see Chapter 4).

Cross-contamination from raw to cooked foods via poor personnel handling practices or through the use of common utensils is also a common factor in many outbreaks. Public education in food safety practices is again important for avoiding such problems. In many cases, the cross-contamination is not a minor occurrence but involves gross contamination from raw meat to ready-to-eat products. Raw meat and poultry should not be stored in contact with or in close proximity to ready-to-eat foods. During preparation, different utensils and working surfaces should be used to allow the preparation of raw and cooked meats separately, or, they must be cleaned and sanitised effectively between each use. Appropriate and high standards of personnel handling practices also need to be employed to prevent the hands or clothing from becoming the vectors of contaminants from raw to ready-to-eat foods.

Poor temperature control (temperature abuse) of foods is often found to be a factor contributing to many foodborne outbreaks, particularly involving *Salmonella*. Clearly, the presence of the viable organism on ready-to-eat foods is not an acceptable situation, but outbreaks are often made worse by an inadequate storage temperature of cooked and ready-to-eat foods in the period prior to consumption allowing multiplication of any *Salmonella* present, sometimes to very high levels. *Salmonella* is no different to many infectious disease agents and although *Salmonella* can cause illness at very low levels in some foods, it is evident that the more cells there are present, the more likely they are to cause disease and the more severe the illness is likely to be.

The simple measures of cooling hot foods quickly and then storing them under good refrigeration conditions could significantly reduce the number of foodborne infections associated with organisms like *Salmonella*. One of the main problems facing home or semi-professional caterers arises because they are often required to cook foods in quantities that exceed their refrigeration capacity. As a consequence, food is often left under exposed ambient conditions, which can provide opportunities for microbial contamination and growth. General advice for maintaining the safety of cooked food is to either keep it hot ($>63°C$) or cool it down within two hours and store in refrigerated ($<8°C$) conditions.

The majority of outbreaks such as those described in these examples could be avoided by taking the simplest of precautions, i.e. avoiding raw egg consumption, particularly by vulnerable groups, cooking foods effectively, avoiding all forms and sources of cross-contamination and then either holding and eating them hot or cooling them quickly followed by proper refrigeration. Consistent application of these precautions by all those preparing or handling food for others to consume would significantly reduce the number of cases of foodborne illness caused by *Salmonella*.

3

FACTORS AFFECTING GROWTH AND SURVIVAL OF *SALMONELLA*

GENERAL

It is clear that, from time to time, *Salmonella* should be expected as contaminants of raw foods, particularly raw meats, but also vegetable crops that may be at risk from contamination by human and animal faecal matter, and appropriate control measures must be instituted in the use of such food materials.

As it is impossible for many raw foods to be produced free from *Salmonella* at source, it is important to establish animal husbandry and crop agricultural regimes that can make a positive contribution to minimising the frequency and level of contamination of these primary raw materials by *Salmonella*. At farm level, effective treatment of animal wastes and slurries by composting of solids or aeration of liquid wastes can achieve useful reductions in pathogen levels. For example, *Salmonella* have been shown to be reduced by >99% of the initial number in slurry by aeration for 2–5 weeks (Heinonen-Tanski *et al.*, 1998). Minimising the levels of *Salmonella* from animal wastes 'cycled' through the soil and water environments can only be of benefit across the full range of primary food production areas.

The important role of good hygienic practices for controlling *Salmonella* throughout animal husbandry, slaughtering processes and the further processing of raw meat and raw meat products has long been well understood (if not always implemented reliably) and was well addressed by a World Health Organisation Expert Committee on Salmonellosis Control (World Health Organisation, 1988).

The introduction and consistent operation of hygienic practices in animal husbandry and crop agricultural systems must be aimed at minimising the faecal contamination load on raw material foods reaching food-processing plants. Thereafter, raw material handling and processing procedures need to be structured and operated to minimise any multiplication of *Salmonella* and prevent cross-contamination of the organisms from animal or crop raw material foods to processed foods or equipment. The low infective doses reported in outbreaks involving some processed high fat content foods, e.g. chocolate and cheese, serve to highlight the necessity for food processors to ensure the maintenance of high quality standards in the design and operation of food processes. This can be achieved effectively through the application of thorough hazard analysis, critical control point (HACCP) based assessments of each process.

In addition to the need for attention to the detail of cleaning and hygiene procedures aimed at reducing the potential for cross-contamination to foods and food contact surfaces and any growth of *Salmonella* in environmental niches, the treatment and formulation of food products are also important for controlling any residual *Salmonella* that may be present.

Within food production processes, a variety of physico-chemical factors, used either singly or in combination, can be effective in controlling the survival and growth of *Salmonella* during processing and also in the finished food product.

TEMPERATURE

The temperature conditions allowing growth of *Salmonella* and for destroying the organism in foods of differing formulation have been the subject of research for over 100 years. In 1904, it was reported that *Salmonella* Typhi, Paratyphi and Enteritidis 'were all killed when the milk was heated to 59°C if 10 minutes was taken to heat the milk to that temperature' and in 1908, work was reported indicating that using nine strains of typhoid bacilli, all were killed in milk heated to 60°C and maintained at that temperature for two minutes (Savage, 1912). Some time–temperature equivalent processes required to effect the same levels of destruction of bacteria in milk were published in 1937 and referred to by Wilson (1942) (Table 3.1). Now, we know also that the sensitivity of *Salmonella* to heat depends on the conditions to which the organism has been exposed prior to heat treatment and the constituents of the medium/food in which the organism is heated.

In conditions of low water activity, heat resistance increases whereas in

Table 3.1 Pasteurisation time–temperature equivalents for milk, adapted from Wilson (1942)

Temperature		Time
°C	°F	
60	140.0	63 minutes
63	145.4	16 minutes*
65	149.0	6 minutes
70	158.0	38 seconds
72	161.6	15 seconds†
76	168.8	2.4 seconds

*Current commercial holder (batch) process = 63°C for 30 minutes.
† HTST = High Temperature Short Time (specified now as at least 71.7°C for 15 seconds; this process was not permitted in the UK until 1941).

conditions of low pH, heat resistance is reduced. In high water activity conditions, *S.* Senftenberg (strain 775W) is a rare heat-resistant strain and is often included as one of the serotypes in experiments on the heat resistance of *Salmonella*. As water activity is reduced, the *D* values of normally heat sensitive serotypes/strains of *Salmonella* can approach or may exceed those of *S.* Senftenberg 775W (Goepfert *et al.*, 1970; Gibson, 1973; Corry, 1974). The temperature limits for growth of *Salmonella* and the *D* values for some *Salmonella* serotypes in different substrates are shown in Tables 3.2 and 3.3, respectively. The protective effect of the food substrate environment is clearly demonstrated by the significantly higher *D* values demonstrated in chocolate, i.e. hours, compared to milk

Table 3.2 Limits for the growth of *Salmonella* under otherwise optimal conditions, adapted from International Commission on Microbiological Specifications for Foods (1996)

Parameter (other conditions being optimal)	Minimum	Maximum
Temperature (°C)	5.2 (most serotypes will not grow at <7.0)	46.2
pH	3.8 (most serotypes will not grow below 4.5)	9.5
Water activity	0.94	>0.99

Table 3.3 *D* values for *Salmonella* in some food substrates

Food substrate/conditions	*Salmonella* serotype	Temperature (°C)	*D* value (seconds, unless otherwise stated)	Reference
Milk (sterile, homogenised)	Anatum	68.3	0.46	ICMSF† (1996)
	Binza	68.3	0.52	
	Cubana	68.3	0.28	
	Meleagridis	68.3	0.4	
	New Brunswick	68.3	0.44	
	Senftenberg	68.3	10.0	
	Tennessee	68.3	0.38	
Ground beef	Typhimurium	63	0.36 minutes	
Milk chocolate	Eastbourne	71	4.5 hours	
	Senftenberg	71	4.6 hours	
	Typhimurium	71	6.6 hours	
Milk chocolate	Anatum	71	20 hours	D'Aoust (1989)
Milk chocolate with 2% water added	Anatum	71	4 hours	
Liquid whole egg, pH 8.0	Typhimurium	60	0.55 minutes	
Liquid whole egg, pH 5.5	Typhimurium	60	2.2 minutes	
Liquid whole egg, pH 8.0	Senftenberg 775W	60	1.5 minutes	
Liquid whole egg, pH 5.5	Senftenberg 775W	60	9.5 minutes	
Liquid whole egg	Typhimurium DT104*	55	6.05 minutes	Jung and Beuchat (2000)
	Non-DT104*		8.04 minutes	

(Continued on p. 88.)

Table 3.3 *Continued*

Food substrate/conditions	*Salmonella* serotype	Temperature (°C)	D value (seconds, unless otherwise stated)	Reference
Liquid whole egg +10% salt	Typhimurium DT104* Non-DT104*	55	4.21 minutes 4.73 minutes	Jung and Beuchat (2000)
Liquid egg yolk	Typhimurium DT104* Non-DT104*	55	9.40 minutes 8.03 minutes	
Liquid egg yolk +10% salt	Typhimurium DT104* Non-DT104*	55	9.06 minutes 10.85 minutes	

*Four-strain mixture.
†ICMSF = International Commission on Microbiological Specifications for Foods.

in which D values for most serotypes are measured in fractions of a second at the temperatures indicated in Table 3.3. Results of some work reported by Juneja and Eblen (2000) demonstrated the effect of fat content in beef on the time to achieve a 7 log reduction in a population of an eight-strain mixture of *Salmonella* Typhimurium DT104. For beef with a 7% fat content, this time was 7.07 minutes at 65°C and with 24% fat content, 20.16 minutes at 65°C. This work also took into account the lag time before any death of *Salmonella* occurred in the meat; the lag time was longer in higher fat content beef, which is probably attributable to the protective effect afforded by the fat which has poorer heat transfer properties than aqueous systems.

The type of solute present in a food also affects the D values of *Salmonella*, e.g. the heat resistance measured between 55°C and 75°C of *S.* Typhimurium and *S.* Senftenberg 775W were reported to increase in solutions of sucrose as the water activity was decreased from 0.995 to 0.706. The heat resistance of these serotypes measured at 65°C was higher in sucrose than in glucose, sorbitol or fructose (D'Aoust, 1989).

In an extensive review of studies on the heat resistance of *Salmonella* in a wide range of foods and culture media, Doyle and Mazzotta (2000) used the data from published reports to calculate that a process of 71°C will require 1.2 seconds to inactivate 1 log cycle of *Salmonella* cells (z value 5.3°C). Although the calculation excluded data for *S.* Senftenberg and data relating to heat resistance following pre-stress applied to cells, the derived process indicates the considerable safety margin built in to the normal milk pasteurisation process, i.e. well in excess of a 10 log cycle reduction in *Salmonella*.

The heat resistance of *Salmonella* can also be increased by heating to sublethal temperatures ($c.$ 50°C) prior to exposure to lethal temperatures, e.g. the time taken to achieve a 1000-fold reduction of a population of *S.* Thompson at 60°C in whole liquid egg approximately doubled after the organisms had been exposed to 48°C for 30 minutes (Mossel *et al.*, 1995). Mackey and Derrick (1986) reported increases in inactivation times up to seven-fold for cells of *S.* Typhimurium pre-incubated at 48°C for 30 minutes prior to heating at temperatures up to 59°C. A similar effect occurs when organisms are heated slowly to the lethal temperature rather than rapidly. This is possibly due, at least in part, to the rapid production of heat shock proteins by the organism during the pre-heating treatment, these helping to confer a higher than normal heat resistance on the organism (Mackey and Derrick, 1990). Changes triggered in the cell membrane composition of heat-stressed *Salmonella* increase the membrane's resistance to heat damage (D'Aoust, 1997). These responses could be of importance to the safety of processes in which ingredients may be held at sublethal temperatures prior to pasteurisation or where the heating is very slow, as in the case of cooked, bulk meats.

Studies of the D values of acid-shocked and non-shocked cells of four-strain mixtures of Typhimurium DT104 and non-DT104 reported by Jung and Beuchat (2000), demonstrated no differences in their heat tolerance at 55°C in liquid egg products. The D values of acid-shocked cells were approximately half that of non-shocked, control organisms; Table 3.3 shows the D values they obtained for non-acid shocked cells.

In other experimental work using broth cultures, high levels (10^8 cfu/ml) of viable competitive microflora have been found to protect a population of *S.* Typhimurium from heat (55°C), increasing the D value for exponential-phase *Salmonella* (10^5 cfu/ml) from 0.4 minutes to 2.09 minutes and also, from freeze injury (Aldsworth *et al.*, 1998). Protection is believed to be afforded by the rapid reduction in levels of dissolved oxygen caused by the high population of competitive organisms present; this reduces oxidative damage to the exponentially growing *Salmonella* bringing the level of resistance of these cells up to that of stationary-phase cells. Other studies using *S.* Enteritidis found the measured heat resistance was up to eight times greater when cells were grown, heated and recovered in anaerobic conditions rather than aerobically (George *et al.*, 1998). The implications of such findings have yet to be considered for food and food production processes but there are many foods in which high levels of microflora are normally present and many other foods that have naturally or artificially reduced oxygen environments within them.

In any hazard analysis of food production processes relying on heat to reduce the level of hazard from *Salmonella*, the effect of the physico-chemical composition of all components of the food and any prior sub-lethal treatment, e.g. heat, must be taken into account when determining the appropriate heat process to be used.

Although variable among strains and dependent on substrate, few *Salmonella* serotypes will grow at temperatures below 7°C, e.g. the minimum reported temperature for the growth of *S.* Heidelberg is 5.3°C and *S.* Typhimurium is 6.2°C (Advisory Committee on the Microbiological Safety of Food, 1992) and the growth rates of all *Salmonella* are greatly reduced at temperatures <15°C. There are reports, summarised by

D'Aoust (1991), of growth of *Salmonella* in shell eggs at 4°C and in minced meat and chicken parts at 2°C but some of these have yet to be confirmed other than by observation of growth on microbiological media (International Commission on Microbiological Specifications for Foods, 1996). Other work has shown *S.* Enteritidis PT4 and *S.* Typhimurium DT104 to increase in biomass in milk, chicken and microbiological media at 4°C due to filament formation rather than cell division (Phillips *et al.*, 1998; ILSI, 2000). Studies with these same strains have demonstrated a similar filamentation, some filaments at least 200 µm long, at low water activities (0.93–0.98) (Mattick *et al.*, 2000). The public health implications of this phenomenon are not known. Confirmation of growth of *Salmonella* in foods at temperatures <5°C whether by cell division or filamentation (while the implications of this have yet to be determined) should be of great concern as effective refrigeration, i.e. temperatures <8°C, is a key control for food safety.

Preservation of food by freezing, particularly at temperatures of −18°C or less, is by no means a reliable method of eliminating the hazard of *Salmonella* if it is present in the food, although some reduction in numbers may be expected particularly during the freezing process itself in the temperature range near freezing, i.e. 0 to −10°C. Indeed, although there is strain-to-strain variation in survival capability, studies of a number of *Salmonella* strains indicate they survive for prolonged periods under freezing conditions (Table 3.4) and some food types afford particular protection to bacteria during freezing and frozen storage, e.g. liquid egg, raw meat and fish (International Commission on Microbiological Specifications for Foods, 1996). Foods intended for freezing therefore should be produced under no less appropriate conditions than foods for chilled storage and freezing should never be regarded or used as a method for making contaminated food 'safe'.

Table 3.4   Survival of *Salmonella* under frozen storage conditions, adapted from D'Aoust (1989)

Serotype	Food	Temperature (minus °C)	Survival time
Enteritidis	Poultry	18	4 months
Choleraesuis	Minced beef	18	4 months
Typhimurium	Chow mein	25	9 months
Enteritidis and Typhimurium	Ice cream	23	7 years
Salmonella spp.	Snails	20	8 years

pH, WATER ACTIVITY AND OTHER FACTORS

In addition to the application of good hygienic practices and well-controlled process temperatures (heat or chill), other physico-chemical characteristics of some specific food types can contribute to the control of the growth of any *Salmonella* that may be present in the food.

Table 3.2 indicates the growth-limiting parameters of pH and water activity for *Salmonella*. For products in which the pH becomes non-optimal (usually acidic) either as a natural result of the manufacturing process, e.g. cheese and fermented meats, or by the direct addition of and mixing with an acidic component, e.g. oil and vinegar (acetic acid) in salad dressings, the pH will contribute to the control of bacterial growth including that of any *Salmonella* which may be present. Organism survival depends on the type of acid present and other physico-chemical conditions prevailing.

When organic acids (acetic, lactic, citric, propionic, etc.) are used as preservatives in food, it is important to ensure the correct concentration of un-dissociated acid (which is responsible for the antimicrobial activity of the acid) is available for bacterial growth inhibition. The proportion of un-dissociated acid present varies with pH (Table 3.5) so this must be taken into account when determining the amount of total acid required at a specific pH to yield a particular concentration of un-dissociated acid. At neutral pH, most organic acids will have a limited effect on the growth of *Salmonella*.

Table 3.5 Percentage of total organic acid un-dissociated at some different pH values, adapted from International Commission on Microbiological Specifications for Foods (1980)

Organic acid	pH value			
	4	5	6	7
Acetic acid	84.5	34.9	5.1	0.54
Citric acid	18.9	0.41	0.006	< 0.001
Lactic acid	39.2	6.05	0.64	0.064

Chung and Goepfert (1970) reported the minimum pH at which *Salmonella* (one or more serotypes of *S.* Anatum, *S.* Tennessee or *S.* Senftenberg) would initiate growth in otherwise optimal microbiological growth medium in which the pH was adjusted with different acidulants to be 4.05 (citric acid), 4.2 (gluconic acid), 4.3 (malic acid), 4.4 (lactic acid), 4.7

(glutaric acid) and 5.4 (acetic acid). At pH values not allowing cell multiplication, cells died and the rate of death was most rapid at the higher incubation temperatures used, i.e. 43°C. Jung and Beuchat (2000) reported a study of the survival of four strains of *S.* Typhimurium DT104 and four strains of non-DT104 *S.* Typhimurium, again using a microbiological growth medium in which the pH was adjusted with different organic acids. They confirmed that at a given pH, acetic acid was the most inhibitory acidulant of those tested. However, in contrast to the results of Chung and Goepfert (1970), at pH 4.0, citric acid was found to be more inhibitory than malic acid and viable cells were still detected in the medium adjusted to pH 3.7 with either of these acids. Jung and Beuchat (2000) concluded from their own work that there was probably little or no difference in the tolerance of cells of DT104 and non-DT104 in the medium acidified with the same acidulant at a given pH.

Membré *et al.* (1997) examined the effects of storage temperature (15–35°C), pH (4.5–6.5), glucose level (1–4% weight/volume) and citric acid level (0.05–0.1% weight/volume) in reduced calorie mayonnaise inoculated with a strain of *S.* Typhimurium isolated from egg products. They reported that whatever conditions were applied, the viable population reduced, but only after a period of time ranging from 11–85 days, thus representing a potential food safety risk, particularly in relation to the use of raw eggs for making these products.

It is clear, that in any hazard analysis of a food production process, due consideration must be given to the potential for acid tolerance to allow *Salmonella* survival. In addition, although *Salmonella* are able to grow over a fairly wide pH range, even if only slowly at the upper and lower limits for growth, death will occur outside this range, but not all cells will die immediately, and slowly declining populations mean that some cells may survive for long periods in acidic products.

Outbreaks of salmonellosis implicating acidic products are witness to the hazard presented by *Salmonella* to such products. Samples of apple cider (juice) taken from the homes of ill persons and the cider production unit during investigation of an outbreak of *S.* Typhimurium were found to have a pH in the range 3.4–3.9 (mean 3.6) (Anon, 1975). The pH of unpasteurised orange juice involved in an outbreak in 1995 in Orlando, Florida, USA, was reported to range from 4.1–4.5 (Cook *et al.*, 1998). Subsequent to this outbreak, studies on the survival of *Salmonella* (four serotypes) inoculated into orange juice and stored at 0°C showed that they could be detected up to 27 days at pH 3.5, 46 days at pH 3.8, 60 days at pH 4.1 and 73 days at pH 4.4 (Parish *et al.*, 1997). Previously reported experiments indicated that from an inoculation level of 10^6 viable cells of *Salmonella* per ml, it took 27 days for four decimal reductions to occur in orange juice of pH 3.0-3.1 held at 5°C (Cook *et al.*, 1998).

Heat treatments applied to products with a sub-optimal pH for *Salmonella* growth and/or containing an organic acid as a preservative are expected to be more effective than the same heat treatment applied to a product at the optimum pH for the organism.

The water activity of a food product is lowered by the addition of sodium chloride, sugars and/or other solutes. The higher the concentration of the solute, the lower the water activity of the product (Table 3.6). For products in which the water activity approaches or is lower than the minimum for growth of *Salmonella*, growth of the organism will be prevented or minimised provided the water activity does not increase during the life of the product, e.g. by mixing the product with other foods of higher water activity or by allowing condensation to affect the product.

Table 3.6 Water activity at various concentrations of sodium chloride or sucrose, adapted from International Commission on Microbiological Specifications for Foods (1980)

Water activity (a_w)	Sodium chloride (%, w/w)	Sucrose (%, w/w) (°Brix)
1.000	0	0
0.99	1.74	15.45
0.98	3.43	26.07
0.96	6.57	39.66
0.94	9.38	48.22
0.92	11.90	54.36
0.90	14.18	58.45
0.88	16.28	62.77
0.86	18.18	65.63

Outbreaks of salmonellosis associated with food products of low water activity, e.g. some fermented meat products, hard cheese, peanut butter, chocolate, dried milk and cereal products and food ingredients such as black pepper and desiccated coconut, clearly indicate the very real hazard presented by *Salmonella* to a wide range of foods and food materials in which they may not actually grow. Indeed, the numbers of any *Salmonella* present may well decline, but sufficient cell numbers to cause illness can and do survive for long time periods. *Salmonella* are reported to survive for more than a year in chocolate, black pepper, and peanut butter; all foods with a low water activity (International Commission on Microbiological Specifications for Foods, 1996). Table 3.7 shows the survival times found in various foods for which low water activity is regarded as a significant controlling factor for food safety. Tamminga (1979) reported that the reduction time in both *S.* Eastbourne and *S.* Typhimurium was slower in milk chocolate (negative from a low inoculum after 14 months and 6 months, respectively) than in bitter chocolate (negative from a low inoculum after 6 months and 34 days, respectively). When milk chocolate was prepared from contaminated milk powder, survival times for organisms increased; *S.* Eastbourne was still detectable after 19 months and test results for *S.* Typhimurium were negative after 15 months. Survival times measured in months are common and survival for years has been recorded, particularly for chocolate products. Low water activities may prevent *Salmonella* from multiplying, but long-term survival presents a hazard to a wide variety of foods relying on low water activity for safety and stability.

A range of other chemical and treatment systems have been explored for minimising the presence and controlling the growth of *Salmonella* in foods and food processes. In the agricultural sector, produce washing and irrigation systems incorporating free (available) chlorine maintained at levels of > 100 ppm can help reduce but not eliminate *Salmonella* on fruit, salad vegetables and sprouted seeds (Jacquette *et al.*, 1996; Beuchat *et al.*, 1998).

In view of the significance of raw meats and poultry products as sources of *Salmonella* in cases of human infection, a great deal of work has been carried out on methods for decontaminating these materials, mainly at the slaughterhouse and butchery stages. Corry *et al.* (1995) provided a useful review of techniques that are in current use and those which may find application in the future following more research into efficacy and practicality. Table 3.8 summarises various approaches to decontaminating raw meat and poultry during primary processing that have been explored either alone or in combination; some are already in use commercially, e.g. steam for beef carcasses and tri-sodium phosphate treatment of poultry carcasses.

Mokgatia *et al.* (1998) reported that some strains of *Salmonella* isolated from poultry being processed through an abattoir were able to grow in the presence of 72 ppm hypochlorous acid and that this level was only bacteriostatic against 50% of the 20 isolates examined. They advocate the use of higher concentrations of hypochlorous acid to ensure all cells are killed in addition to the careful control of cross-contamination.

Table 3.7 Some examples of the survival of *Salmonella* in foods of low water activity

Food	*Salmonella* serotype(s)	Inoculum level per g	Water activity (a_w)	Survival time	Reference
Cheddar cheese	Typhimurium PT10			8 months at 5°C	D'Aoust *et al.* (1985)
Dried milk products	Naturally contaminated with 3 serotypes			Up to 10 months	Ray *et al.* (1971a)
Pasta	Infantis, Typhimurium		Moisture content c. 12% (w/w)	Up to 12 months	Rayman *et al.* (1979)
Milk chocolate	Eastbourne	10^8	0.41	>9 months at 20°C	Tamminga *et al.* (1976)
		10^5	0.38	9 months at 20°C	
Bitter chocolate	Eastbourne	10^7	0.51	9 months at 20°C	
		10^5	0.44	76 days at 20°C	
Peanut butter	5 serotypes: Agona Enteritidis Michigan Montevideo Typhimurium	$5.68 \log_{10}$ cfu	0.2–0.33	Up to 24 weeks held at 5°C or 21°C	Burnett *et al.* (2000)
		$1.51 \log_{10}$ cfu	0.2–0.33	Up to 24 weeks held at 5°C	
				Up to 6 weeks held at 21°C	
Paprika powder	Multiple serotypes			>8 months	Lehmacher *et al.* (1995)

Table 3.8   Some examples of decontamination treatments examined for use on raw meat and poultry carcasses and meats, adapted from Corry *et al.* (1995), Bolder (1997), Tamblyn and Conner (1997), Delmore *et al.* (2000) and Natrajan and Sheldon (2000a and b)

Physical treatment	Chemical treatment
Water – different temperatures, rinse, spray, waterfall	Chlorine (sodium hypochlorite) – spray or immersion
Steam with or without vacuum	Organic acids (acetic, citric, gluconic, lactic, malic, tartaric) in dips or sprays
Irradiation	
Pulsed electric fields	Inorganic phosphates (tri-sodium phosphate, polyphosphates)
Ultrasonic energy	Organic preservatives (benzoates, sorbates)
Oscillating magnetic field pulses	
High pressure	Oxidising agents (ozone, hydrogen peroxide)
High intensity visible light	Transdermal compounds – sodium lauryl sulphate, ethanol, sorbitan monolaurate, dimethyl sulfoxide in combination with organic acids
Ultraviolet light	
Microwaves	
	Nisin-based formulations in edible films applied to poultry skin surfaces or on packaging films

There has been increasing interest in, and study of, the antimicrobial activity of herbs and spices as well as their essential oils. Highly variable antimicrobial activities have been reported (International Commission on Microbiological Specifications for Foods, 1998; Hammer *et al.*, 1999) and, in foods, the concentration of any plant oil is often too low to exert any tangible antimicrobial effect although they may provide some synergistic activity together with other antimicrobial systems present. It is prudent however, not to rely on the presence of herbs, spices and/or their oils for providing control of the growth of any *Salmonella* that may be present in the food.

Any physical or chemical treatment proposed to be applied in any food production situation with the aim of reducing/eliminating specific pathogens such as *Salmonella* should be thoroughly evaluated including, if necessary, the use of challenge studies to ensure the level of reduction sought can be consistently attained.

It will always be important to procure and use raw materials of high initial microbiological quality and, where appropriate, these should be bought against a microbiological specification requiring the non-detection of *Salmonella* in 25 g or more of the material. Thorough and structured hazard analysis of all aspects of a food production process should ensure that processes will be sufficiently well controlled to prevent additional contamination and survival and/or growth during the product's shelf life of any *Salmonella* already present. Attention to the detail of such analysis and controls will greatly assist the prevention of outbreaks of human salmonellosis.

4

INDUSTRY FOCUS: CONTROL OF
SALMONELLA

INTRODUCTION

Salmonella is probably the most researched of the foodborne pathogens due to its recognition in food poisoning outbreaks that have occurred over many decades. There are few foods that have not been implicated in outbreaks of salmonellosis and examination of these outbreaks can provide valuable information regarding the factors important for controlling the organism.

As one might expect, it is usually a failure in food production/manufacturing control systems that allows the organism to gain entry and/or survive to cause an outbreak of illness. Many such failures occur in the control of very basic areas such as cross-contamination, cooking processes, personal hygiene and food storage temperature. In such cases, proper hazard analysis and implementation of relevant controls at the critical points identified could have prevented the outbreaks, provided the control systems were operated consistently correctly. It is strongly recommended that all persons involved in the primary production, processing and sale of food adopt a hazard analysis approach to food safety considering all relevant pathogens, including *Salmonella*.

To help focus attention on the products representing the greatest concern in relation to *Salmonella* and the areas requiring greatest management control, a series of questions can be applied to each food process/product. Processes and products can be reviewed against the key questions to identify the level of concern that *Salmonella* may represent (Table 4.1). As a guide to answering these questions, some familiar products in different commodity groups are given as examples (Table 4.2).

Table 4.1 How much of a concern does your product represent?

Question?	Yes	No
Is *Salmonella* expected to be present in the raw material and, if so, will it be present at a high (e.g. > 1%) incidence or low incidence?	(High/Low)	
Is the raw material of animal origin or exposed to contamination from animal sources, e.g. wastes/contaminated irrigation water?		
Is the raw material of poultry origin or exposed to contamination from poultry sources, e.g. wastes/contaminated water supplies?		
Will the organism be destroyed or reduced to an acceptable level by any of the processing stages?		
Will the product be exposed to any post-process contamination?		
Can the normal process and product conditions allow the organism to survive and grow?		
Will the product be subjected to a process by the customer that will destroy *Salmonella*?		

After answering each of the questions in Table 4.1, the product can be assessed against the profiles given in Table 4.3 to determine the level of concern that may be associated with the product. Having done this, the key process areas requiring greatest attention for control of the hazard can be determined (Table 4.4).

Every process and product will differ from those presented in the tables; therefore, the tables should be used for guidance purposes only. Complete understanding of the hazard and controls can only be gained by applying a full hazard analysis to each process and product. In addition, it is important to note that even processes and products which are rated as being of very low concern in relation to *Salmonella* may still be capable of causing outbreaks if the controls inherent in the normal manufacture of these products are not applied correctly. In fact, significant hazards to food safety are presented by complacent management teams who believe that their product is safe because of historical precedence or from a food production team lacking the necessary skills and training in safe food

Table 4.2 Examples of key process stages where *Salmonella* may represent a hazard in different foods*

Product	Product examples	Raw material contamination	Raw material of animal (non-avian) origin	Raw material of poultry origin	Reduction process	Destruction process	Post-process contamination	Process allows growth	Consumer cidal process
Dairy products									
Raw milk ripened soft cheese	Raw milk Brie, Camembert	Low/High	Yes	No	No	No	Yes	Yes	No
Raw milk hard cheese	Raw milk Cheddar, Parmesan	Low/High	Yes	No	Yes	No	Yes	No	No
Pasteurised milk ripened soft cheese	Brie, Camembert	Low/High	Yes	No	Yes	Yes	Yes	Yes	No
Pasteurised milk hard cheese	Edam, Cheddar, Cheshire	Low/High	Yes	No	Yes	Yes	Yes	No	No
Pasteurised milk fermented products	Yoghurt, fromage frais, cottage cheese	Low/High	Yes	No	Yes	Yes	No	No	No
Spray dried milk powder	Infant dried milk	Low/High	Yes	No	Yes	Yes	Yes	No	No

See p. 104 for key to table.

(Continued on p. 102.)

Table 4.2 *Continued*

Product	Product examples	Raw material contamination	Raw material of animal (non-avian) origin	Raw material of poultry origin	Reduction process	Destruction process	Post-process contamination	Process allows growth	Consumer cidal process
Meat products									
Raw meat	Pork, lamb, beef	High	Yes	No	No	No	Yes	No	Yes
Raw poultry	Chicken, duck, turkey	High	No	Yes	No	No	Yes	No	Yes
Fermented and dry-cured meat	Salami, Parma ham	High	Yes	Yes/No	Yes	No	Yes	No	No
Cooked meat (hermetically sealed)	Whole hams, chub pâté	High	Yes	Yes/No	Yes	Yes	No	No	No
Cooked meat (sliced or whole)	Ham, beef, frankfurters, pâté	High	Yes	No	Yes	Yes	Yes	No	No
Cooked poultry (sliced or whole)	Chicken, turkey, pâté	High	No	Yes	Yes	Yes	Yes	No	No
Fish and shellfish									
Raw fish/shellfish	Cod, mussels	Low	Yes	No	No	No	Yes	No	Yes

See p.104 for key to table.

Table 4.2 *Continued*

Product	Product examples	Raw material contamination	Raw material of animal (non-avian) origin	Raw material of poultry origin	Reduction process	Destruction process	Post-process contamination	Process allows growth	Consumer cidal process
Raw fish/shellfish, consumed raw	Oysters, sushi	Low	Yes	No	No	No	Yes	No	No
Cold smoked fish, consumed raw	Smoked salmon and smoked trout	Low	Yes	No	No	No	Yes	No	No
Cooked fish/shellfish/fish pâté	Prawns, crab	Low	Yes	No	Yes	Yes	Yes	No	No
Salad and vegetables									
Raw vegetables	Potato, beans, peas	Low	Yes/No	Yes/No	No	No	Yes	No	Yes
Raw salads	Lettuce, spring onions, celery	Low	Yes/No	Yes/No	No	No	Yes	No	No
Prepared salads	Salad mix	Low	Yes/No	Yes/No	Yes	No	Yes	No	No
Sprouted vegetables	Bean sprouts, cress, alfalfa sprouts	Low	Yes/No	Yes/No	Yes	No	Yes	Yes	No

See p. 104 for key to table.

(Continued on p. 104.)

Table 4.2 *Continued*

Product	Product examples	Raw material contamination	Raw material of animal (non-avian) origin	Raw material of poultry origin	Reduction process	Destruction process	Post-process contamination	Process allows growth	Consumer cidal process
Other foods									
Chocolate	Chocolate bars and coated confectionery	Low/High	Yes/No	No	Yes	Yes/No	Yes	No	No
Raw eggs	Hen eggs, duck eggs	Low/High	No	Yes	No	No	Yes	No	Yes/No
Fruit juice – unpasteurised	Orange juice, apple juice	Low/High	Yes/No	Yes/No	Yes	No	Yes	No	No

*Information given is for guidance only and may not be appropriate for individual circumstances. It is recommended that proper hazard analysis is carried out for every process and product to identify where controls must be implemented to minimise the hazard from *Salmonella*.

Raw material contamination: Is *Salmonella* expected to be present in the raw material, if so, will it be present at high or low incidence?

Raw material of animal origin: Is the raw material of animal origin or exposed to sources of animal contamination, e.g. wastes?

Raw material of poultry origin: Is the raw material of poultry origin, i.e. chicken, turkey, duck, or exposed to sources of poultry contamination, e.g. wastes?

Reduction or destruction process: Will the organism be reduced to an acceptable level or destroyed by any of the processing stages?

Post-process contamination: Will the product be exposed to any post-process contamination?

Process allows growth: Are the normal processing conditions suitable for the growth of *Salmonella*, if present?

Consumer cidal process: Will the product normally be subjected to a process by the customer that will destroy *Salmonella*?

Table 4.3 Categories of concern*

Level of concern	Product examples	Raw material contamination	Raw material of animal (non-avian) origin	Raw material of poultry origin	Reduction process	Destruction process	Post-process contamination	Process allows growth	Consumer cidal process
Category 1: Highest	Raw milk ripened soft cheese	Low/High	Yes	No	No	No	Yes	Yes	No
	Sprouted vegetables, bean sprouts, alfalfa	Low	Yes/No	Yes/No	Yes	No	Yes	Yes	No
Category 2: High	Raw eggs	Low/High	No	Yes	No	No	Yes	No	Yes/No
	Salami, dry-cured ham	High	Yes	Yes/No	Yes	No	Yes	No	No
	Raw milk hard cheese	Low/High	Yes	No	Yes	No	Yes	No	No
	Chocolate products	Low/High	Yes/No	No	Yes	Yes/No	Yes	No	No
	Infant dried milk powder	Low/High	Yes	No	Yes	Yes	Yes	No	No
	Raw poultry/ comminuted poultry products, burgers	High	No	Yes	No	No	Yes	No	Yes

(Continued on p. 106.)

See p. 107 for key to table.

Table 4.3 *Continued*

Level of concern	Product examples	Raw material contamination	Raw material of animal (non-avian) origin	Raw material of poultry origin	Reduction process	Destruction process	Post-process contamination	Process allows growth	Consumer cidal process
Category 2: High	Cooked poultry (sliced or whole)	High	No	Yes	Yes	Yes	Yes	No	No
	Fresh pressed fruit juice – unpasteurised	Low/High	Yes/No	Yes/No	Yes	No	Yes	No	No
Category 3: Medium	Prepared salads	Low	Yes/No	Yes/No	Yes	No	Yes	No	No
	Oysters, sushi, smoked salmon	Low	Yes	No	No	No	Yes	No	No
	Pasteurised milk ripened cheese, Brie	Low/High	Yes	No	Yes	Yes	Yes	Yes	No
	Cooked meat (sliced or whole) and cooked shellfish	Low/High	Yes	No	Yes	Yes	Yes	No	No
	Pasteurised milk hard cheese	Low/High	Yes	No	Yes	Yes	Yes	No	No
	Raw red meats	High	Yes	No	No	No	Yes	No	Yes

Table 4.3 *Continued*

Level of concern	Product examples	Raw material contamination	Raw material of animal (non-avian) origin	Raw material of poultry origin	Reduction process	Destruction process	Post-process contamination	Process allows growth	Consumer cidal process
Category 4: Low	Raw fish and shellfish, cod, plaice, mussels	Low	Yes	No	No	No	Yes	No	Yes
Category 5: Lowest	Chub pâté, products cooked in pack	Low/High	Yes/No	Yes/No	Yes	Yes	No	No	No

*Information given is for guidance only and may not be appropriate for individual circumstances. It is recommended that proper hazard analysis is carried out for every process and product to identify where controls must be implemented to minimise the hazard from *Salmonella*.

Highest concern: Where *Salmonella* could be present due to raw material contamination or as a post-process contaminant and where the process allows survival and growth and the product is ready to eat.

High concern: Where *Salmonella* could be present due to raw material contamination or as a post-process contaminant and where the process reduces but does not destroy the organism and the product is ready to eat with minimal consumer processing *or* it is a poultry product *or* a baby food.

Medium concern: Where *Salmonella* may be present in the raw material or as a post-process contaminant and where the process usually achieves a reduction of the organism (low contamination level of raw material) or destruction of the organism (high contamination level of raw material) or a consumer cidal process is applied.

Low concern: Where *Salmonella* may be present in the raw material (low contamination level) and the process applied destroys the organism but it is subject to post-process handling and a consumer cidal process is applied.

Lowest concern: Where *Salmonella* may be present in the raw material but the process applied destroys the organism and it cannot re-contaminate the product.

Table 4.4 Process stages where control of *Salmonella* is critical (based on the categories of concern)*

Level of concern	Product examples	Raw material control	Reduction process	Destruction process	Post-process contamination	Process conditions	Consumer issues
Category 1: Highest	Raw milk ripened soft cheese	Yes			Yes		Yes
	Sprouted vegetables, bean sprouts, alfalfa	Yes	Yes		Yes	Yes	Yes
Category 2: High	Raw eggs	Yes			Yes		Yes
	Salami, dry-cured ham	Yes	Yes		Yes		
	Raw milk hard cheese	Yes	Yes		Yes		
	Chocolate products	Yes		Yes	Yes		
	Infant dried milk powder	Yes		Yes	Yes		Yes
	Raw poultry/comminuted poultry products, burgers	Yes			Yes		Yes
	Cooked poultry (sliced or whole)	Yes		Yes	Yes		
	Fresh pressed fruit juice – unpasteurised	Yes	Yes		Yes		
Category 3: Medium	Prepared salads	Yes	Yes		Yes		
	Oysters, sushi, smoked salmon	Yes			Yes		Yes

Table 4.4 *Continued*

Level of concern	Product examples	Raw material control	Reduction process	Destruction process	Post-process contamination	Process conditions	Consumer issues
Category 3: Medium	Pasteurised milk ripened cheese, Brie	Yes		Yes	Yes		
	Cooked meat (sliced or whole) and cooked shellfish	Yes		Yes	Yes		
	Pasteurised hard cheese	Yes		Yes	Yes		
	Raw red meat	Yes			Yes		Yes
Category 4: Low	Raw fish and shellfish	Yes			Yes		Yes
Category 5: Lowest	Chub pâté, products cooked in pack			Yes			

*Information given is for guidance only and may not be appropriate for individual circumstances. It is recommended that proper hazard analysis is carried out for every process and product to identify where controls must be implemented to minimise the hazard from *Salmonella*.

Highest concern: Where *Salmonella* could be present due to raw material contamination or as a post-process contaminant and where the process allows survival and growth and the product is ready to eat.

High concern: Where *Salmonella* could be present due to raw material contamination or as a post-process contaminant and where the process reduces but does not destroy the organism and the product is ready to eat with minimal consumer processing *or* it is a poultry product *or* a baby food.

Medium concern: Where *Salmonella* may be present in the raw material or as a post-process contaminant and where the process usually achieves a reduction of the organism (low contamination level of raw material) or destruction of the organism (high contamination level of raw material) or a consumer cidal process is applied.

Low concern: Where *Salmonella* may be present in the raw material (low contamination level) and the process applied destroys the organism but it is subject to post-process handling and a consumer cidal process is applied.

Lowest concern: Where *Salmonella* may be present in the raw material but the process applied destroys the organism and it cannot re-contaminate the product.

manufacture. Food products are generally made safe or unsafe to eat by human intervention.

The products of highest concern with regard to *Salmonella* are those where the organism may be present, even in low numbers, in the raw material, where no process exists to reduce or eliminate it and where it may survive or even grow during the process or in the finished product which is consumed without any further processing. Products such as raw-milk, mould-ripened, soft cheeses, sprouted salad vegetables and eggs (consumed raw or used in dishes that are uncooked) would fall within this category and, indeed, these products have caused many outbreaks of salmonellosis. History has also shown that raw products heavily con-taminated with *Salmonella* are also a significant potential cause of out-breaks because of the difficulties of ensuring that caterers (including restaurants) and consumers handle and cook them safely.

SALAMI AND RAW, DRY-CURED MEAT PRODUCTS

Raw fermented meat products, i.e. salami and dry-cured meats, are traditional products, the manufacturing processes for which have developed over many centuries. Salamis, in particular, have been implicated in a number of salmonellosis outbreaks over the years (Cowden *et al.*, 1989; Sauer *et al.*, 1997; Pontello *et al.*, 1998). Although less frequently, raw, dry-cured meat has also caused salmonellosis outbreaks (González-Hevia *et al.*, 1996)

Salamis and raw, dry-cured meats are manufactured and consumed throughout the world and most countries have unique recipes and manufacturing processes that deliver the wide variety of products seen on the market today. Traditional salamis are perhaps most synonymous with mainland European countries such as Germany, Denmark, Italy and Spain, i.e. German salami, Danish salami, etc. Raw, dry-cured meat products command the premium end of this market and include products such as Parma ham, a famous food export of Italy. The products are consumed by all age groups and certain varieties are particularly attractive to children, e.g. 'snack' salami.

The method of manufacture of these products differs markedly between salami and raw, dry-cured meats but also there are a considerable number of process variations for each product type.

Description of process

Salamis are manufactured from raw meat which is cut into small pieces and mixed, in a bowl chopping process, with fat, salt, herbs, spices and usually a fermentable sugar (Figure 4.1). The preservatives sodium nitrite and sometimes sodium nitrate are added to the mixture. Many salami processors employ the use of lactic acid bacterial starter cultures to aid the subsequent fermentation, but a number of products are manufactured without these. The meat mixture is stuffed into natural or artificial casings and then tied at each end to form a sausage shape. The individual salamis are attached to racks on a frame support and this is placed in a fermentation chamber.

The chamber is usually under both temperature and humidity control, although this will vary depending on the size and complexity of the manufacturing operation. Fermentation conditions vary quite markedly across different processes although the greatest difference tends to be between processes used to produce European-style and American-style

Process Stage	Consideration
Animal husbandry ↓	Health Cleanliness
Animal slaughter and processing ↓	Hygiene Temperature
Meat transport, delivery and storage ↓	Hygiene Temperature Intake quality control
Meat trimming ↓	Hygiene
Bowl chop with fat, herbs, spices, salt, nitrite, sugar (and starter culture) ↓	Distribution of additives Starter culture activity Hygiene
Stuff into casing ↓	Hygiene Casing quality assurance
Tying ↓	Hygiene
Fermentation ↓	Hygiene Temperature Humidity Time Acidity development/pH profile
Drying ↓	Hygiene Temperature Humidity Time Moisture loss
Slicing, where applicable ↓	Hygiene
Packing ↓	Hygiene
Storage and distribution ↓	
Retail slicing, where applicable, and sale	Hygiene

Figure 4.1 Process flow diagram and technical considerations for a typical salami.

salamis. In the former, the product is fermented for longer periods (*c.* 5 days) at lower temperatures 20–30°C, whereas the latter employ shorter fermentations at higher temperature (> 30°C). In both cases, the objective of this stage is to achieve an effective fermentation, brought about by the growth of lactic acid bacteria, which results in a rapid reduction in pH due to the production of organic acids. The pH after fermentation usually approaches or is slightly below pH 5.0, although this can vary, particularly for products not employing the use of starter bacteria. During fermentation, some moisture is also lost from the product resulting in slight drying, but the main drying stage occurs after fermentation.

Following fermentation, the salamis undergo a drying process, sometimes in a separate chamber in which temperature and humidity are controlled. Some products may be smoked before drying. Again, the drying stage of the process can vary markedly in different salami processes, some decreasing in temperature and some gradually increasing in temperature as the product becomes progressively drier. Nevertheless, the aim of this stage of the process is always the same, to achieve moisture loss, and, in doing so, create a product that is microbiologically stable; after all, this process has been used and developed over many centuries to preserve meat that would otherwise have succumbed to microbial spoilage.

Drying stages can vary from a week to several months and during the longer drying processes mould growth often occurs on the exterior surface of the product imparting a distinctive ripened flavour. Mould growth often causes an increase in pH, particularly at the surface of the product. The mould is usually brushed or washed off the product prior to sale.

Finished products usually have a low water activity ranging from 0.94 to below 0.85 and a low pH (4.5–5.0). However, mould-ripened varieties may have pH values which vary from < 5.0 in the centre to 5.5–7.0 towards the surface.

Finished products are sold whole for slicing on delicatessen counters or they may be pre-sliced at the factory and sold as pre-packaged units. Some 'snack' salamis may be packed in a modified atmosphere or pasteurised in pack.

Raw, dry-cured meats are manufactured from whole anatomical pieces of meat. The meat pieces are trimmed to the desired amount of fat and then salted by the application of dry salt to the external surfaces. The salt mixture may contain sodium nitrite and nitrate and also some herbs and spices. The salted meat is placed into refrigerators and kept at a low temperature (*c.* 3°C) for several days or weeks (Figure 4.2). The meat is removed on a number of occasions, turned and fresh salt rubbed onto the meat. During this stage, the high salt content and low temperature select a dominant lactic acid bacterial population.

Process Stage	Consideration
Animal husbandry ↓	Health Cleanliness
Animal slaughter and processing ↓	Hygiene Temperature
Meat transport, delivery and storage ↓	Hygiene Temperature Intake quality control
Meat trimming ↓	Hygiene
Salting and storage Curing ↓	Distribution of salt Hygiene Temperature Time
Drying ↓	Hygiene Temperature Time Moisture loss
Slicing, where applicable ↓	Hygiene
Packing ↓	Hygiene
Storage and distribution ↓	
Retail slicing, where applicable, and sale	Hygiene

Figure 4.2 Process flow diagram and technical considerations for a typical raw, dry-cured meat product.

After storage, the meat is taken out and placed into a drying chamber at temperatures of <15°C where it is kept for several months extending up to a year or more for some products. Some products may be smoked prior to drying. During drying, a mild bacterial fermentation takes place on the surface as the product dries to a very low water activity (*c.* 0.9 to < 0.85). The pH of the finished product is usually between pH 5 and 6.

The product is sold as a whole piece of meat for slicing on the delicatessen counter or may be sliced at the factory where it is often vacuum or modified atmosphere packed prior to selling as pre-packs.

Because the combination of organic acids, low pH, high aqueous salt content and low water activity of many salami and raw, dry-cured meat products will not allow the growth of bacterial pathogens and spoilage organisms, long product shelf lives (several months) are often allocated to these products. Due to the inherent stability of the finished products, they are displayed under ambient temperature storage conditions in many countries. However, it is common practice in the UK for these products to be stored and displayed under refrigeration conditions (< 8°C).

Raw material issues and control

The main raw material for the manufacture of salami and raw, dry-cured meats is the meat itself and the most common meat used is pork. For salami, pork may be used solely or it may be mixed with beef and it is becoming increasingly common for chicken or turkey to be used as ingredients. As well as pork, beef is also used for the production of some raw, dry-cured meat products.

The biggest microbiological hazard to these products clearly arises from the potential contamination of the raw meat with pathogens including *Salmonella*. It is well known that raw meat may be contaminated by a wide variety of *Salmonella* serotypes and the incidence of contamination can be extremely high, depending on the meat species being used. Surveys of the incidence of *Salmonella* in pork have been reviewed by D'Aoust (1989) and show a wide variation from 0.4% in a Swedish survey to 76.3% in a Dutch survey with the average incidence being 16.2%. These figures relate to surveys carried out many years ago and, with improved practices in animal rearing and processing, this incidence is now, on average, much lower, i.e. *c.* 5%. For example, in a survey of 1420 samples of caecal material (1 g samples) from pigs at a Canadian abattoir, the incidence of *Salmonella* was found to be 5.2% (Letellier *et al.*, 1999). A survey of pig caecal contents in the UK (10 g samples) revealed a *Salmonella* incidence of 23%, with *S.* Typhimurium and *S.* Derby being the dominant serotypes (Davies, 2000). Carcass contamination on slaughtered pigs at UK abattoirs was also surveyed; this demonstrated a 5.3% (135/2509 carcass swab samples, 0.1 m^2) incidence of *Salmonella* with

S. Typhimurium (2.1%) and *S.* Derby (1.6%) again being the predominant serotypes (Davies, 2000). Nevertheless, the incidence can vary significantly and monitoring the microbiological quality of the incoming material is an important element of the raw ingredient quality assurance programme for these products.

Incidence of *Salmonella* in other raw meat species also varies, with the highest being found in poultry meat, where incidence regularly exceeds 50% (D'Aoust, 1989; Bryan and Doyle, 1995). A survey of raw broiler chickens (25 g samples of leg, thigh and breast meat and skin) on sale in Korea found an incidence of 25.9% (Chang, 2000). The reported incidence of *Salmonella* on fresh poultry carcasses in the UK decreased from 54% in 1987 to 33% in 1994 (Table 1.6) and is approximately 10–20% today but the incidence in frozen poultry is often found to be nearly double that of fresh birds.

Salmonella contamination of beef occurs at a much lower frequency than poultry. A survey of cattle presented at slaughter in Great Britain found a carriage rate of *Salmonella* (1 g caecal samples) of 0.22% (2/891) (Evans, 2000). In a major study of 3780 samples (100 cm^2 swabs) from beef carcasses in the USA, contamination with *Salmonella* ranged from an average incidence of 0.8–5.0% in pre-eviscerated steer-heifers to 2.1–15.5% in cow-bulls (Sofos *et al.*, 1999). After carcass washing and chilling, the average incidence decreased to 0–2.5% in steer-heifers and 0–4.4% in cow-bulls. This is consistent with a previous study of 2089 samples of steer and heifer carcasses which was part of the USA nation-wide beef baseline data collection programme carried out between October 1992 and September 1993 (Anon, 1994). The average incidence in this survey was found to be 1%, and of the positive carcasses the mean level of contamination was determined to be 0.1 MPN/cm^2 and the highest still only < 1 MPN/cm^2.

The microbiological integrity of the raw meat is therefore of utmost importance to the ultimate safety of these products. It is not possible to preclude the possibility of *Salmonella* being present in raw meat but it is possible to ensure that poor quality meat, with unacceptably high levels of contamination, are not routinely being supplied. This can be achieved by effective raw material quality assurance programmes that involve auditing suppliers and selecting only those suppliers operating high standards of hygienic processing in the abattoir and in meat cutting operations. This can be supported by routine monitoring of meats received for microbial indicators of faecal contamination such as *E. coli*, high levels (> 10/cm^2) of which may give some cause for concern. Microbiological examination may also usefully include *Salmonella*, to ensure that the incidence is not high. Although, its presence may be anticipated from time to time, high incidence, e.g. > 5% in pork, > 1% in beef, should be of concern and action taken by the supplier to reduce incidence.

Raw meat is usually used quickly (48–72 hours) when received and stored under chilled conditions. However, meat may also be received frozen and then defrosted prior to use. It is important to ensure that defrost conditions do not allow the proliferation of any microbial contaminants that may be present on the meat and this should therefore be conducted under well-controlled, chilled conditions.

Raw meat for salami and raw, dry-cured meat production is further cut and trimmed at the processing site. It is important to recognise that this stage, even though involving the handling of raw meat, has significant potential to spread contaminants that may be localised on one carcass to many others. This can occur from knives, meat trays, conveyors, hands or other product contact and environmental sources. It is therefore essential to employ effective cleaning and sanitisation procedures in these areas to prevent such occurrences.

Other raw ingredients should not be overlooked for their potential to introduce microbial contaminants into the raw material mixture, particularly for salami. For example, different herbs and spices may be added, and many of these are known sources of *Salmonella*. Such material bought in untreated should be routinely monitored for the presence of *Salmonella* but, where possible, these are best sourced as heat processed ingredients. The use of natural casings for salami also needs careful consideration. Natural casings are derived from the intestines of animals and are subject to washing, salting and drying to remove/reduce contaminants, so the potential to introduce enteric pathogens from this source must be considered during the hazard analysis conducted on the safety of these products. It is clear that a number of the raw materials and the meat in particular will introduce *Salmonella* into salami mixes from time to time and, therefore, the subsequent process must be designed and operated with this hazard in mind.

Process issues and control

Fermented meat

Salami is produced by bowl chopping the meat together with all of the other ingredients. This process effectively distributes microbial contaminants initially present on the external surfaces of the meat onto the surfaces of all the resulting meat pieces throughout the mix. When the mixture is then forced into casings, the contaminants will be distributed throughout the salami stick.

Significant opportunity exists for cross-contamination between salami batches during the bowl chopping stage and effective cleaning and sanitisation of the bowl chopper is essential to ensure that microbial contamination does not build up. Enteric pathogens are usually present as contaminants on the raw meat but can be readily transferred throughout individual batches and on to successive batches of salami by the operation of poor hygienic practices at the bowl chopping stage. Bowl choppers are notoriously difficult pieces of equipment to clean effectively and they should be properly stripped down, cleaned and sanitised. The efficacy of cleaning should be checked by routine visual inspection supported by microbiological swabbing for indicators such as Enterobacteriaceae. The use of rapid indicators of cleaning efficacy such as ATP-bioluminescence tests can be a useful verification check of effective cleaning.

It is easy for personnel working with raw meat to mistakenly believe that good standards of hygiene and cleaning are not necessary and important because of the subsequent processing that may take place. However, with processes such as salami manufacture, assumptions of this kind are very dangerous indeed as poor cleaning can result in the extensive spread of contaminants that can jeopardise the safety of a larger number of successive production batches. In addition, if persistently poor cleaning leads to high levels of contaminants being introduced in the meat mixture, these levels may exceed the capacity of the subsequent process to reduce them to a safe level.

The bowl chopping stage is also important in relation to achieving effective distribution of the ingredients, many of which are crucial to the subsequent safety of the process and final product. Nitrite (>75 ppm ingoing), salt (2–5% ingoing), fermentable sugar and starter cultures, if used, need to be properly distributed throughout the meat mixture in order to ensure they can contribute uniformly to the intrinsic conditions necessary to facilitate a safe process.

Once bowl chopped, the salami paste, which may contain meat pieces of varying size, is stuffed into casings. This forms a sausage shape that is closed at both ends using metal crimps or tied using plastic or natural string. The size of the salami is dictated by the diameter of the casing and the points at which it is tied and can range from 10 cm to >40 cm in length and from 1–10 cm in width. The diameter of the salami is an important factor in determining the length of time necessary for the subsequent fermentation and drying processes in order to achieve both a safe and desirable end product.

Once formed, the salami sticks are usually hung on a frame in large numbers and then placed in a fermentation room of controlled temperature and humidity. The salamis should be hung with sufficient space between them to allow free circulation of air. It is during this stage that the fermentation process begins. General factors that should be considered for ensuring effective process control during the fermentation include the temperature of the ingoing salami paste, the environmental temperature and humidity.

Salamis are fermented for 2–5 days at warm temperatures, between 20°C and 30°C in Europe and above 30°C in the USA. These temperatures are essential for facilitating the rapid growth of the lactic acid bacteria present that cause the necessary fermentation activity in the salami mixture. Salamis may be manufactured without the use of starter culture bacteria, and whilst it is possible to achieve effective fermentation using the naturally derived microflora, it is generally believed that the use of starter cultures makes the fermentation process more uniform and reliable. Nevertheless, whether using starter cultures or not at this stage of the process, it is vital to ensure that an active fermentation, with concomitant production of organic acids and decreases in pH, occurs throughout each product and the batch. At the temperatures employed during fermentation, a significant amount of moisture loss also occurs (up to 15%) and these changes are ultimately very important to the safety of the final product in relation to the reduction of any *Salmonella* that may have been present in the raw ingredients.

The pH of the salami typically decreases from an initial pH of < 6.0 to below pH 5.0 within 48 hours and many achieve a pH as low as 4.5. The combination of active lactic acid bacterial growth, production of organic acids, depression in pH and decreasing moisture levels together with the presence of salt and nitrite, all act to preclude the growth of *Salmonella* in the early stages of fermentation. If fermentation did not occur, or occurred slowly, it is possible for *Salmonella* to grow in the conditions present in the initial meat mixture and therefore active fermentation is essential for the control of *Salmonella*. The rate of pH decrease and/or acidity increase are usually monitored for each fermentation process. The normal profiles of these parameters for individual processes should be established and complied with for every successive batch of product.

After fermentation, the salami is transferred to drying rooms where the temperature is usually decreased to *c.* 15°C or less and the humidity is controlled. It is stored here for several weeks or months to allow continued moisture loss from the product. Some small diameter salamis are only fermented and dried for very short periods because moisture loss occurs very quickly due to the large surface area to volume ratio of the product. Although they ferment and dry quickly, it does not follow that the rate of pathogen elimination is equally fast. In fact, outbreaks of salmonellosis attributed to 'snack' salami demonstrate that the short fermentation and drying times do not always result in sufficient reduction in contaminating pathogens.

A number of challenge test studies have been conducted to determine the survival of *Salmonella* during the manufacturing process for salami.

Smith *et al.* (1975) assessed the survival of a number of *Salmonella* strains during the manufacture of pepperoni by three processes. Cultures of *S.* Dublin and *S.* Typhimurium were inoculated into a beef–pork meat mixture containing 3% sodium chloride and 1.2 g sodium nitrate/kg meat. Pepperoni was made with the addition of a proprietary starter culture mix (*Pediococcus cerevisiae* and *Lactobacillus plantarum*), without added starter but after ageing the meat used at 5°C for 10 days, and without added starter culture using un-aged meat. The different batches were hand mixed and then stuffed into 55-mm fibrous casings prior to fermenting at 35°C (85% relative humidity) and then drying at 12°C (60–65% relative humidity).

Table 4.5 summarises the results obtained from these experiments. Clearly, the use of starter cultures had a significantly enhanced effect on the destruction of *Salmonella* in the salami process, although an effective fermentation brought about by natural microflora in properly aged meat also contributed to a good reduction (100 to 100 000-fold) of the organism by the end of the manufacturing process. In addition, the importance of extended drying times for the effective destruction of *Salmonella* is evident from the long time required to achieve a significant reduction even in a properly fermented product.

The researchers also examined the efficacy of a heating stage introduced at the end of fermentation, but prior to drying, on the destruction of the *Salmonella* strains. Pepperoni processed as described above were taken immediately after fermentation and heated in a smoking room until the centre temperature in the salami reached 60°C; they were then immediately cooled and subjected to the drying process at 12°C for 14 days. In

Table 4.5 Survival of *Salmonella* during pepperoni manufacturing processes, adapted from Smith *et al.* (1975)

Pepperoni type	Day	pH	a_w	Level of microorganism/g sausage	
				S. Dublin	*S.* Typhimurium
With added starter culture*	Fermentation				
	0	6.1		4.5×10^4	8.2×10^3
	1	4.5		8.8×10^3	4.5×10^2
	Drying†				
	1	4.5	0.960	3.3×10^4	1.8×10^3
	8	4.5	0.935	2.2×10^4	1.2×10^2
	15	4.6	0.885	6.1×10^2	6.0×10^1
	22	4.6	0.875	1.6×10^2	9.3
	29	4.6	0.828	2.2×10^2	0.03
	43	4.6	0.798	2.4×10^1	<0.03
With natural microflora (aged meat)‡	Fermentation				
	0	6.2		3.5×10^4	1.0×10^4
	1	5.5		4.2×10^4	1.3×10^3
	2	5.1	0.978	8.0×10^3	2.5×10^3
	Drying†				
	7	4.9	0.941	2.0×10^4	3.0×10^2
	14	5.1	0.885	1.5×10^4	1.2×10^2
	21	5.0	0.878	7.5×10^3	4.6×10^1
	28	5.0	0.853	6.0×10^2	2.1
	42	5.0	0.805	1.0×10^2	0.04
With un-aged meat and no starter culture‡	Fermentation				
	0	6.1		4.1×10^4	8.9×10^3
	1	6.1		3.4×10^6	2.4×10^5
	2	6.0	0.993	7.3×10^5	2.2×10^4
	Drying†				
	7	5.8	0.973	2.1×10^5	1.2×10^5
	14	5.8	0.896	6.0×10^5	1.8×10^4
	21	5.8	0.878	4.1×10^5	4.5×10^3
	28	5.7	0.855	1.2×10^5	2.8
	42	5.7	0.828	1.3×10^6	0.07

* 1 day fermentation at 35°C, 85% relative humidity.
† 12°C, 60–65% relative humidity.
‡ 2 day fermentation at 35°C, 85% relative humidity.

pepperoni made with starter culture, after natural ageing of the meat (no added culture) and with no ageing and no added culture, the pH prior to heating was 4.6, 4.8 and 5.8, respectively and the level of *S*. Dublin was found to be 2.0×10^2/g, 3.2×10^3/g and 1.6×10^2/g, respectively. These levels decreased to < 0.03/g in all cases after heating to an internal temperature of 60°C and remained so after the 14 days drying period (Table 4.6).

Table 4.6 Effect of heating to an internal temperature of 60°C on the survival of *Salmonella* in pepperoni after the fermentation stage, adapted from Smith *et al.* (1975)

	Pepperoni type					
	With added starter culture*		With natural microflora† (aged meat)		With un-aged meat and no starter culture†	
	pH	Level of *S*. Dublin/g	pH	Level of *S*. Dublin/g	pH	Level of *S*. Dublin/g
Before heating	4.6	2.0×10^2	4.8	3.2×10^3	5.8	1.6×10^2
After heating	4.7	< 0.03	4.9	< 0.03	6.1	< 0.03
After 14 days drying‡		< 0.03		< 0.03		<0.03

* 1 day fermentation at 35°C, 85% relative humidity.
† 2 day fermentation at 35°C, 85% relative humidity.
‡ 12°C, 60-65% relative humidity.

Ellajosyula *et al.* (1998) studied the survival of *S*. Typhimurium in Lebanon bologna, a fermented and heat-treated meat product that undergoes no drying stage. They described the product as being routinely manufactured from beef to which salt, sodium nitrite and starter cultures are added. It is fermented for 24–30 hours at 70-100°F (21.1-37.8°C) during which time the pH of the meat decreases from *c*. 6.0 to 5.2-4.7. The temperature of the product is then raised to 110-120°F (43.3-48.9°C) for 11-35 hours. Ellajosyula *et al.* (1998) used a Lebanon bologna mix containing beef, salt (3.5%), potassium nitrate (12 ppm), sodium nitrite (200 ppm), a commercial spice formulation and a commercial starter culture containing *Pediococcus acidilactici*, *Lactobacillus plantarum* and *Micrococcus* spp. The mix was inoculated with *c*. 10^8 cfu/g *S*. Typhimurium and then fermented at 80°F (26.7°C) for the first 12 hours and then held at 100°F (37.8°C) until the pH reached 5.2 or 4.7. The mix was then heated to 110°F (43.3°C) (within 0.5 hour) or to 115°F (46.1°C) in 5.5 hours or to 120°F (48.9°C) in 5 more hours, i.e. 10.5 hours, and held at these temperatures for defined periods of time. Fermentation alone to pH 5.2 or 4.7 without any subsequent heat treatment achieved little if any reduction in

S. Typhimurium. The reduction in the product fermented to pH 5.2 was < 0.5 log unit and when the pH after fermentation achieved 4.7 the reduction was only slightly greater, achieving just above 1 log unit reduction. Conditions that gave a > 7 log reduction included fermentation to pH 5.2 or 4.7 followed by heating to 110°F (43.3°C) for 20 hours, 115°F (46.1°C) for 10 hours or gradual heating to 120°F (48.9°C) over a 10.5 hour period after fermentation.

Goepfert and Chung (1970) studied the survival of *Salmonella* in a cured sausage emulsion stored at 30°C in a beaker in the presence and absence of starter culture. The mixture contained ground beef (100 g), glucose (1.5 g), sodium chloride (3.0 g) and sodium nitrite (0.01 g). In the presence of starter culture, *S.* Typhimurium decreased from 1.1×10^4/g at time 0 to 2.4×10^2/g after 24 hours (Table 4.7). Similar reductions were observed for a range of other *Salmonella* strains (Table 4.8). When no starter culture was present the levels increased from 39/g to 1.1×10^5/g within 14 hours.

Table 4.7 Survival of *S.* Typhimurium in a cured sausage batter* with and without starter culture stored in a beaker at 30°C, adapted from Goepfert and Chung (1970)

Time (hours)	*S.* Typhimurium (per g)	
	Starter culture present	No starter culture
0	1.1×10^4	3.9×10^1
6	1.5×10^3	2.4×10^2
14	—	1.1×10^5
24	2.4×10^2	—

*Formulation of batter mixture: beef (100 g), glucose (1.5 g), sodium chloride (3.0 g) and sodium nitrite (0.01 g).
— = no examination.

When inoculated into a commercial batter mix for a Thuringer sausage (a fermented meat product containing salt, nitrite and starter culture, which is cooked after the primary fermentation), *S.* Typhimurium decreased from 3.5×10^5/g to 3.5×10^4/g after a fermentation stage in which the pH decreased to pH 4.85 (initial pH 6.0) (Goepfert and Chung, 1970). The levels further reduced to 2.4×10^3/g after bringing the fermented Thuringer sausage up to a temperature of 50°C and holding it at this temperature for 8 minutes.

Studies to determine the heat treatment required to destroy *S.* Typhimurium in Thuringer sausage were conducted using sausage fermented at

Table 4.8 Survival of *Salmonella* during fermentation of a cured sausage batter*
at 30°C in a beaker, adapted from Goepfert and Chung (1970)

Time (hours)	S. Thompson	S. Choleraesuis	S. Infantis	S. Anatum
0	1.1×10^3	4.6×10^4	3.9×10^4	1.1×10^5
12	1.5×10^3	4.3×10^1	2.4×10^4	4.6×10^3
24	2.4×10^2	9.1×10^0	4.6×10^3	4.6×10^2

* Formulation of batter mixture: beef (100 g), glucose (1.5 g), sodium chloride (3.0 g) and
sodium nitrite (0.01 g).

30°C to a pH of 4.8 and then elevating the temperature of the sausage to
three different heating temperatures (46°C, 49°C and 52°C). The time
taken to elevate the temperature of the sausage ranged from 45–60 minutes, after which the products were held at the appropriate temperatures
for several hours. Heating the fermented sausage to 46°C for up to 5 hours
achieved less than 1 log unit reduction, whereas more than a 3 log
reduction was achieved after 5 hours at 49°C and in excess of a 4 log
reduction was achieved after just 1 hour at 52°C (Table 4.9).

Table 4.9 Reduction of *S.* Typhimurium in fermented Thuringer sausage heated
at different temperatures, adapted from Goepfert and Chung (1970)

	Levels of *S.* Typhimurium (per g)					
	pH	Temperature (46°C)	pH	Temperature (49°C)	pH	Temperature (52°C)
Pre-fermentation	6.1	2.1×10^4	6.0	2.4×10^5	6.0	2.4×10^4
Post-fermentation	4.8	2.4×10^3	4.8	1.1×10^4	4.8	2.4×10^3
Post heating for 1 hour*		4.6×10^2		2.1×10^3		<0.03
Post heating for 3 hours		4.6×10^2		4.6×10^2		<0.03
Post heating for 5 hours		2.4×10^2		2.4×10^0		<0.03

* Time to heat sausage to holding temperature ranged from 45–60 minutes.

The enhanced effect of low pH on the destruction of *Salmonella* seen in
the previous study was clearly demonstrated when the same heat
destruction study was conducted in low-acid Thuringer sausage fermented to only pH 5.2–5.4 and then subjected to processing at 46°C, 49°C
and 52°C for 1–5 hours (Table 4.10).

It should be remembered that at such low temperatures employed in the
heating of these products, it is likely that significant variation may occur in
the level of destruction of different *Salmonella* serotypes. Heat sensitivity
is affected by both the pH and water activity of products (see Chapter 3).
It is recommended that manufacturers, relying on heating at low

Table 4.10 Reduction of *S*. Typhimurium in 'low-acid' (pH 5.2–5.4) Thuringer sausage heated to different temperatures, adapted from Goepfert and Chung (1970)

	Levels of *S*. Typhimurium (per g)		
	Temperature		
	(46°C)	(49°C)	(52°C)
Pre-heating	2.4×10^4	1.1×10^4	1.1×10^4
Post heating for 1 hour*	4.6×10^3	1.1×10^4	2.4×10^1
Post heating for 3 hours	1.1×10^5	2.4×10^3	<0.03
Post heating for 5 hours	1.1×10^6	2.4×10^2	<0.03

*Time to heat sausage to holding temperature ranged from 45–60 minutes.

temperature to destroy *Salmonella* in these or any similar products, should conduct properly designed process validation studies using research facilities and including a representative range of *Salmonella* serotypes in the commercial product in question.

It is clear from these studies that well-controlled fermentation processes make a useful contribution to microbiological safety but that the ultimate safety of fermented meat products in respect of the hazard of *Salmonella* is also dependent on the effective application of processes following fermentation whether drying, heating, or both.

Raw, dry-cured meats

Raw, dry-cured meats are manufactured from whole anatomical meat pieces. Therefore, the majority of microbial contaminants are located on the external surfaces of the meat. The meat is butchered, trimmed of excess fat and then dry salt is rubbed over the surfaces. The meat is placed in layers in a container/vessel and stored in a cold-room at *c*. 3°C for long periods, usually exceeding seven days. The meat is regularly removed, turned, re-salted and then replaced. During this stage of the process, salt equilibration occurs in the meat which, together with the low storage temperature, allows the selective development of a lactic acid bacterial population. Although using salami, the work already described by Smith *et al.* (1975) clearly demonstrated the importance of holding meat at low temperature for long enough time to allow the development, prior to the fermentation stage, of an appropriate natural lactic microflora for the production of these types of product.

After the period of salting, the meat is moved to a drying chamber in which temperature and humidity is controlled. Here, it is dried at 10–15°C for several weeks or months, extending up to a year for some products. Some processors dry and smoke products concurrently and some progressively elevate temperature as the product gradually dries out.

The microbial contaminants at the surface of the meat are exposed to harsh physico-chemical conditions arising from the surface application of dry salt, so they are directly exposed to the inhibitory effects of the salt, the drying process and the competitive effects associated with the growth of other microorganisms present. Any population of *Salmonella* present will be greatly reduced under these conditions.

Other products

A variety of other fermented and raw, dried meats are manufactured throughout the world using variations of the processes described above. Beef jerky is a widely consumed product in the USA, often made in the home using kitchen-top dehydrators. Beef jerky has been implicated in outbreaks of salmonellosis. One outbreak involving commercially made product in the USA in 1995 affected 93 people. *Salmonella* Typhimurium, *S.* Montevideo and *S.* Kentucky were isolated from stool samples of those affected and also from samples of beef jerky (Crespin *et al.*, 1995). The implicated jerky was produced by placing partially frozen strips of raw beef in a drying room (60°C) for 3 hours and then holding the meat at 46°C for 19 hours prior to sale to the public.

This product is traditionally manufactured by cutting raw beef into small strips and then marinating these, under refrigeration conditions for several hours, in a sauce comprising smoke flavour, pepper, garlic, salt and a variety of other spices. It is then placed in a dehydrator and dried at temperatures of *c.* 60°C or more for 10–12 hours. Although beef jerky has been implicated in outbreaks of salmonellosis, studies have demonstrated that providing the temperatures during the drying stage are sufficiently high then significant and sufficient reduction in *Salmonella* will occur to preclude outbreaks.

Harrison and Harrison (1996) inoculated *S.* Typhimurium into strips of beef and stored them in a marinade containing 60 ml soy sauce, 15 ml Worcestershire sauce, 0.6 g pepper, 1.25 g garlic powder, 1.5 g onion powder and 4.35 g hickory smoke-flavoured salt per 900 g meat. They were stored at 4°C for 1 hour and then placed in a dehydrator operated at 60°C for 10 hours. The dehydrator raised the internal temperature of the meat strips to 60°C within the first 4–5 hours and the moisture content of the meat decreased from 69.9% to 23.8% during the process. Levels of *S*. Typhimurium decreased by 3.1 log units after 3 hours of drying, by over 5 log units after 6 hours and by a total of 5.5 log units after the full 10 hours drying period. Such levels of reduction are sufficient to make beef with normally low incidence and levels of contamination safe from the hazard of *Salmonella.*

Final product issues and control

Salamis and raw, dry-cured meats are usually considered to be ambient stable products. With the exception of products that undergo very short drying stages, most of these products have water activities below 0.94 and usually between 0.90 and 0.85. The pH of salamis ranges from 4.5 to 5.5 in the centre and up to 6–7.0 on the surface of mould-ripened varieties. The pH of raw, dry-cured meat is usually *c.* 5.5.

Both product types are sold either as whole products to retail stores for slicing on the delicatessen counter, or they may be sliced at the factory or an intermediary unit and pre-packed for retail sale. Clearly, opportunities exist for the introduction of contaminants during the slicing and packing stages. Thorough cleaning and sanitisation of slicing machines and effective control of personal hygiene and handling practices are important for preventing pathogens from gaining access to these products.

Many salamis have post-process additions made to them for the purposes of flavour or decoration. The most visible of these are peppered salamis that have a coating of gelatine added to the outside followed by rolling in crushed peppercorns. Post-process additions can introduce significant further sources of potential microbial contamination.

Gelatine handling practices need to be carefully controlled as this is often held molten in tanks for extended periods during use. The use of acidified gelatine or holding the gelatine at elevated temperature ($>60°C$) will preclude the growth of *Salmonella* but this may cause functional deterioration of the gelatine. As a minimum, the gelatine should be regularly disposed of and the holding vessel and associated equipment should be cleaned and sanitised frequently. The pepper also constitutes a significant potential hazard, as this material will occasionally be contaminated with *Salmonella* through the growing, harvesting and drying conditions employed in its production. Peppercorns can be steam pasteurised and re-dried prior to use and many spice manufacturers employ such systems to decontaminate this material. Wherever possible, pasteurised peppercorns should be used for the production of peppered salami products.

Products are sold either 'loose' or pre-packed in vacuum or modified atmosphere pack conditions. The latter packaging conditions are usually used to preclude the growth of moulds and also to prevent further drying. In most European countries, salamis are stored under ambient conditions, but in the UK it is common practice to store these products in refrigerated conditions. Whilst pathogenic bacteria should clearly not be present at this stage, it is known that further reduction in such contaminants does occur in salami during ambient storage due to the harsh conditions present. Ng *et al.* (1997) studied the survival of a combination of three *Salmonella* serotypes (*S.* Typhimurium, *S.* Enteritidis and *S.* Choleraesuis) inoculated onto finished dry-cured hams subsequently stored at 2°C and 25°C for up to 28 days. The six different products examined had chemical compositions ranging as follows: nitrite (0.95–9.8 ppm), salt (3.8–9.4%), moisture (48.8–63.8%) and pH (6.0–6.4). Reductions in *Salmonella* over the 28-day storage period ranged from 2 $\log_{10}$ cfu/g to $>5 \log_{10}$ cfu/g at 25°C and from 0.6 $\log_{10}$ cfu/g to 3.4 $\log_{10}$ cfu/g at 2°C. The mean difference between the reduction in *Salmonella* populations when stored at 2°C and 25°C at the end of the 28 days was 2.2 $\log_{10}$ cfu/g. This is further evidence of the enhanced destruction of *Salmonella* that can be obtained in these products when stored at ambient temperature and the enhanced survival of the organism in products stored under refrigerated conditions.

Chilled storage of products was cited as a possible contributory reason for the outbreak caused by the consumption of snack salami in the UK (Cowden *et al.*, 1989).

Microbiological surveys of salamis and raw, dry-cured meats have occasionally found *Salmonella* together with indicator organisms/groups such as *E. coli* and Enterobacteriaceae. In a survey of 455 samples of salamis and raw, dry-cured meats on sale in the UK, one sample (25 g) was found to be contaminated with *Salmonella* (Anon, 1997d). The *Salmonella* was identified as *S.* Typhimurium PT29 and it was isolated from a sample of Parma ham. Ninety-seven per cent of samples had Enterobacteriaceae counts of < 10/g and 99% of *E. coli* results met this same standard. The range of counts of *E. coli* and Enterobacteriaceae reported are shown in Table 4.11. It is normal practice to carry out routine microbiological testing on several samples of each salami/raw, dry-cured meat batch, including tests for Enterobacteriaceae, *E. coli* and *Salmonella*.

The safety of fermented and raw, dry-cured meat products is dependent on a wide range of factors that are not entirely understood. In general, the reduction of *Salmonella* is dependent on an active and rapid fermentation brought about by added starter culture or natural microflora together with

Table 4.11 Incidence of Enterobacteriaceae, *E. coli* and *Salmonella* in a survey of salami and raw, dry-cured meat on sale in the UK, adapted from Anon (1997d)

Organism	Level	Description of product						
		Italian salami	German salami	Pepperoni	Country ham	Cervelat	Dried meat	Not known
		Number of samples in range						
Enterobacteriaceae ($\log_{10}$ cfu/g)	<1.00	95	179	57	49	44	3	16
	1.00–1.99	2	2	1	1	1	0	0
	2.00–2.99	0	1	0	0	1	0	1
	3.00–3.99	0	0	0	0	0	0	0
	4.00–4.99	1	0	0	0	0	0	0
	5.00–5.99	0	1	0	0	0	0	0
E. coli ($\log_{10}$ cfu/g)	<1.00	98	182	58	50	46	3	17
	1.00–1.99	0	1	0	0	0	0	0
Salmonella (presence in 25 g)		0	0	0	1*	0	0	0

* Parma ham.

an extended, controlled drying period. Destruction of *Salmonella* is significantly affected by the period of drying and any product not undergoing an effective fermentation and a long drying period (> 3 weeks) should be considered to represent an elevated risk in relation to the survival of *Salmonella* and other enteric pathogens.

It is recommended that products made by processes that have short fermentation and drying times should be either pasteurised, to ensure elimination of *Salmonella*, or the ability of the proposed process to reduce the organism to an acceptable level should be proved by challenge testing using appropriate methods and research facilities. Indeed, because of the variations of processes employed for the manufacture of these products it is recommended that most products of this nature be challenge tested to establish their safety. If insufficient reduction, e.g. < 5 log cfu/g, is achieved then modifications to the process by adding an increased level of starter culture, extending the drying period or introducing a pasteurisation stage should be considered. The outbreaks of salmonellosis associated with these products reinforce the need to understand the processes and factors important in reducing contamination with enteric pathogens and to ensure that necessary controls are routinely adopted and consistently effectively operated.

COOKED, PERISHABLE MEAT PRODUCTS

Cooked meat products are commonly implicated in outbreaks of food-borne disease and salmonellosis outbreaks have been reported implicating most meat species. Cooked meats are consumed throughout the world and are made from almost any animal species, the most common being poultry, beef, pork and lamb. These days, exotic species of animal meat are more noticeable in Western diets including ostrich and kangaroo, although these would not be considered exotic in their native countries. Cooked meats are consumed by all sectors of society, from the very young to the elderly, usually without any further re-heating. Due to their perishable nature, they are usually sold as chilled products in whole meat cuts or slices, e.g. cooked chicken breast, cooked, sliced ham or roast beef.

Description of process

The process involved in the manufacture of cooked meat products (Figure 4.3) varies according to the actual type of product required. In general, whole cuts or comminuted and reformed pieces of raw meat, to which salt or spices may be added by brine injection or by tumbling, are sealed

Process Stage	Consideration
Animal husbandry ↓	Health Cleanliness
Animal slaughter and processing ↓	Hygiene Temperature
Meat transport, delivery and storage ↓	Hygiene Temperature
Comminuted and reformed bulk meats Bowl chopping and addition of other ingredients (spices, herbs, salt, etc.) or	Hygiene Temperature Distribution of ingredients
Whole joints of meat Brine injection (where applicable, e.g. hams and cured meats) ↓	Hygiene Temperature Brine strength Distribution of brine

Figure 4.3 Process flow diagram and technical considerations for a typical cooked, sliced meat. *(Continued on p. 132.)*

Process Stage	Consideration
Cooking ↓	Temperature Time High/low-risk segregation (post cooking)
Blast chilling ↓	Hygiene Temperature Time
Removal from container (where applicable) ↓	Hygiene
Storage ↓	Hygiene Temperature
Roasting/chilling (where applicable) ↓	Temperature Time
Super chilling ↓	Hygiene Temperature
Slicing (where applicable) ↓	Hygiene
Garnishing ↓	Raw material control Hygiene
Packing ↓	Hygiene Temperature
Storage/distribution ↓	Temperature
Retail storage ↓	Hygiene Temperature
Retail slicing (where applicable) ↓	Hygiene
Retail sale ↓	Hygiene Temperature
Consumer	Shelf life Temperature

Figure 4.3 *Continued*

into bags, containers or cans prior to cooking. Products such as whole chicken or chicken portions are usually cooked without sealing into bags, although many are brine injected or coated in flavouring prior to cooking. Also, chicken pieces pre-treated with salt or spices, may be forced into casings to form a large sausage shape which is sealed at both ends prior to cooking and subsequent use for sliced chicken products.

The meat is cooked in an oven operated to achieve a centre temperature in the meat sufficient to destroy enteric pathogens. In the UK, the minimum commercial process recommended for cooking meat is 70°C for two minutes or an equivalent heat process. Ovens vary in design from types in which the meat is held on a static rack and air is circulated within the oven, to rotary ovens where the meat rack is continuously rotated throughout the cook.

The cooking stage is always defined as a critical control point at which the achievement of an adequate cook and the avoidance of cross-contamination from the raw meat to the cooked product are both essential to product safety. It is therefore normal for cooked meat manufacturers to use double-entry ovens; the oven being built into a dividing wall to provide an effective separation between raw and cooked material. Product is loaded into the oven from the raw side, the door closed and the meat cooked. After cooking, the product is removed through the second oven door on the cooked side of the factory thereby precluding the opportunity for raw meat or equipment and personnel from the raw side to contaminate the cooked product. After removal from the oven the meat is blast chilled to quickly achieve temperatures below 5°C and prevent the germination and growth of any surviving spore-forming bacteria.

Meat cooked in bags may be supplied as bulk units for slicing on the retail delicatessen counter or, more commonly, is sliced by the manufacturer for sale as pre-packed units. After cooking, some meats are subjected to a further roasting stage to impart a roasted appearance prior to slicing, whilst others such as whole chicken may be further processed into portions or stripped for supply as meat pieces for products such as ready-meals and sandwiches. Cooked meat products are distributed and sold under refrigerated conditions (<8°C).

The shelf life of cooked meat products is highly variable and depends on the meat type, process employed and finished product characteristics. Uncured cooked meat products subject to post-cooking processing, e.g. stripping, slicing, etc., such as chicken, pork and beef are allocated lives ranging from less than 10 days up to 20 days. This is due to the potential for the presence and growth of contaminating spoilage bacteria and psychrotrophic pathogens such as *Listeria monocytogenes* and non-proteolytic strains of *Clostridium botulinum*. Cooked cured meats such as ham and pâté and those products cooked in their final container, e.g. canned, cured perishable meats, are usually allocated much longer shelf lives ranging generally from 15 days for ham and pâté to >3 months for the canned products. All these products are stored under refrigeration at 5-8°C or less and effective refrigeration (<8°C) should preclude the growth of *Salmonella*. Although some *Salmonella* strains are reportedly capable of growth at temperatures down to 5°C, their growth in cooked meat under effective refrigeration would not normally be expected. It should be noted that refrigerated storage is intended to maintain the safety and quality of these product types by restricting the growth of psychrotrophic pathogens and spoilage microorganisms. Although effective in restricting the growth of *Salmonella*, it is not principally employed for this purpose as clearly, the organism should not be present in the cooked product in the first place.

Raw material issues and control

Cooked meats are generally manufactured using relatively few ingredients, the principle raw material being the meat itself although increasingly these days, added value meat products are being produced using a wider range of raw materials, particularly flavourings, herbs and spices, e.g. barbecue, oriental and spicy flavours, etc. The meat component may be derived from a wide variety of animals but the main species still used are poultry, pork, beef and lamb. Occasionally, raw materials of this nature will be contaminated with *Salmonella*.

The frequency and level of contamination varies markedly between different animal species and, depending on the rearing and slaughtering conditions, can also be significantly different between animals from the same species originating from different supply sources. The incidence of *Salmonella* is often reported to be the highest in raw poultry with reports of incidence throughout the world frequently exceeding 50%. In the UK, it is possible to source poultry from farms and processors regularly achieving an incidence below 10% and in some countries, such as Sweden, industry and government controls have resulted in an incidence of less than 1%. In general, experience has shown that poultry which is intended for direct sale as fresh, raw poultry has a lower incidence of *Salmonella* than frozen birds, raw poultry portions or poultry for further processing. This may be due to the processing conditions employed in the manufacture of frozen birds and the greater opportunity for spread of contamination among bulk delivered birds/portions for further processing. Although relatively frequent in occurrence, the level of *Salmonella* on poultry is not usually high and on most occasions is < 100 per washed carcass.

The incidence of *Salmonella* on other meat types is usually much lower than that found on poultry, although pork can sometimes be contaminated at a very high incidence. On average, the incidence of *Salmonella* in pork rarely exceeds 5% although historical reports have demonstrated incidence above 75% with an average incidence of 16.2% (D'Aoust, 1989). *Salmonella* contamination rates of beef are usually much lower than poultry and pork. It would be unusual to find an incidence of *Salmonella* in beef exceeding 5% and large surveys in the USA and Great Britain have found an average incidence of 1% or below (Anon, 1994; Evans, 2000). The levels of *Salmonella* contamination in pork and beef are usually very low. Studies on beef demonstrated that carcasses positive for *Salmonella* were contaminated with a mean level of 0.1 MPN/cm^2 (Anon, 1994).

As well as differing in the incidence and level of contamination between different meat species, the type of *Salmonella* can also vary with *S.* Enteritidis being more commonly isolated from poultry and *S.* Typhimurium more commonly isolated from beef and pork. Although, in general, the type of *Salmonella* present is of little significance to the effectiveness of the subsequent heat process, some concern has been expressed about the increasing isolation of *Salmonella* serotypes showing antibiotic resistance, e.g. *S.* Typhimurium DT104. However, as most *Salmonella* serotypes do not differ significantly in their thermal resistance, even this new concern is of no significance in relation to the meat cooking processes used.

Effectively managed raw material quality assurance programmes are important in limiting the incidence and particularly the levels of microbial contamination on the incoming meat. Whilst it is difficult to envisage a practical scenario in which the raw material will be totally free from contamination with *Salmonella* it may be possible to considerably reduce the frequency and level at which this hazard enters the processing plant on the incoming meat ingredients. To help achieve this, abattoirs supplying meat should be routinely audited and meat should be subjected to intake (point of receipt) testing for bacterial indicator organisms such as *E. coli* and Enterobacteriaceae, as well as occasional surveys of the incidence of *Salmonella*. The information obtained should be used to discuss with the supplier the ongoing microbiological quality of the material being received and to ensure that poor quality material is not being routinely supplied.

Once received, meat is stored either frozen or chilled. *Salmonella* would not be expected to grow under good refrigeration conditions ($<8°C$) given the short shelf life of most raw meat and poultry and as raw meat is usually stored at temperatures well below this ($0-2°C$, if chilled), increases in any initial contamination levels will not occur. It is important to ensure that subsequent defrosting of all frozen meat is carried out under controlled chilled conditions and not at ambient temperatures as local elevation of temperature at the surface of the meat could allow growth of *Salmonella* and other organisms.

It should not be forgotten that other ingredients are also used in the manufacture of cooked meats, although few have the potential to be so frequently contaminated with *Salmonella* as the raw meat. Herbs and spices together with an increasing range of flavourings are used extensively in the curing mixture injected into or applied in chopping or mixing processes or to the surfaces of the meat. As many of these are supplied as flavour extracts in an inhibitory carrier, e.g. ethanol, glycol, etc., contamination with *Salmonella* will rarely be a concern. However, many brines contain natural flavours or fresh/dried herbs and spices and these can present a potential hazard in respect of *Salmonella* contamination. Herbs and spices should be routinely monitored for the presence of *Salmonella* and even though a heat process is to be employed during the production of the final meat product, levels of contamination from any ingredient should be kept to a minimum.

It should always be remembered that the subsequent heat process is designed to achieve a particular organoleptic quality as well as a certain reduction in initial levels of contaminants; if the levels of microorganisms in the raw ingredients increase, the chances of their survival also increases.

Other ingredients may also be added to the mixture including salt, emulsifiers (di- and polyphosphates or citrates) and sodium or potassium salts of nitrate or nitrite (for cured meats). Handled correctly, there is little potential for these to introduce contaminants such as *Salmonella*.

Process issues and control

Processing of cooked meat begins with the preparation of the raw meat. Beef and pork are usually received as whole carcasses or primal cuts and are butchered and trimmed according to the recipe. Some products are manufactured from whole cuts of meat, although the majority is made up from chopped meat that may be mixed with fat and other ingredients such as herbs and spices during a bowl chopping or mixing stage. These comminuted meats are reformed and packed into casings made usually of plastic or metal. Poultry is often cooked as whole carcasses and may have flavouring added to the outside prior to roasting.

Most large meat pieces have salt added by brine injection. A brine solution, which may include emulsifying salts and preservative agents such as sodium nitrite or nitrate, is injected into the deep muscle of the meat through needle injectors. The meat may then be stored or tumbled for periods of time under controlled temperature conditions to allow the brine to equilibrate throughout the meat before cooking. A similar process is often employed for poultry prior to cooking. Occasionally, salt and preservative agents are added by dry salting the meat and/or tumbling it in a salt solution, but clearly, equilibration of salt will take longer using this more traditional approach. Small meat pieces for use in a reformed meat product may also be tumbled with dry salt or in a brine solution before further processing.

These primary stages of processing prior to cooking can often be overlooked as of little consequence to product safety. However, as the cooking process will only achieve a pre-defined reduction in the level of any pathogens present, any stage that allows pathogens to spread or increase in number can lead to levels of contamination that might exceed the capacity of the heat process to deliver a reduction to safe levels.

Raw meat preparation processes have the potential to allow extensive spread of contaminants from one meat carcass to the entire batch if inadequate standards of hygiene are in operation. Therefore, effective hygiene procedures in the raw meat preparation rooms together with the operation of strictly maintained cleaning and sanitisation regimes of raw meat processing equipment are essential components for ensuring product safety.

One part of the preparation process that should come under particular scrutiny is the brine injection system. Brine solutions are made up and used in a multitude of ways depending on the equipment available and the specific product type required. Most microbial contaminants are located on the skin or surface muscle tissue of the meat and the deep muscle tissue of meat and poultry is generally free from contamination with organisms such as *Salmonella*. It is important to recognise that as brine is injected into the centre of meat pieces, any contaminants in the brine will also be injected into the deepest part of the meat. The physical process of injecting the brine needles into the flesh also has the potential to transfer contaminants from the surface into the deep tissues, although this is likely to be of less significance than direct injection of organisms present in the brine solution itself.

Re-circulatory brine injection systems that collect any brine that flows out of the meat product during injection and transfers it back into the brine tank to be re-injected with fresh brine carry a higher risk of transferring microbial contamination into the muscle tissue than single dose systems. As re-circulatory systems are operated for long periods of time, contaminants can gradually increase in number through concentration, considerably increasing the levels of contaminants that will survive in the brine environment and from here enter the deep muscle tissue. High levels of microorganisms in the brine may significantly elevate the chances of contaminants such as *Salmonella* surviving the subsequent heat process. Re-circulatory systems should be subject to strict controls such as limiting the time that a brine tank is in use before emptying and cleaning it. This is especially important when it is used for poultry carcasses where the loading of bacterial contaminants and *Salmonella* can be extremely high.

Many brine systems are not operated under chill conditions and although the brine strength, which is normally in excess of 20%, should preclude the growth of organisms such as *Salmonella*, the potential for growth of other bacteria such as the toxigenic staphylococci should be considered. In re-circulation systems, the brine in the reservoir readily becomes contaminated with nutrients and organisms from the blood of the meat (and sometimes pieces of meat and fat) and this rich mixture facilitates the survival of vegetative bacterial contaminants. Where lower strength brine solutions are used they may present a more significant hazard, particularly as the brine becomes more nutritious (with blood, etc.) so supporting the growth of some contaminants. In such cases solutions should be chilled or used within a short period.

Although a variety of preservative agents may be added to some meat product types prior to cooking, some of which may affect the growth potential of *Salmonella*, the quantities added have little effect on the survival of the organism and are not considered critical to its control. Nitrite is added to cured meats such as hams and pâté but it is rarely added to cooked beef or poultry products. Salt concentrations in cooked meats are usually lowest in poultry and beef products where levels are generally between 1% and 2% in the aqueous phase. Hams usually have aqueous salt concentrations in excess of 2%. Emulsifiers such as di- and poly-phosphates or citrates may be added to levels of 1–2% in the aqueous phase. Emulsifiers depress the water activity of the product due to their water binding properties but they have little effect on the survival of *Salmonella* in the context of cooked meat products.

The critical stage in the manufacture of cooked meats is obviously the cooking stage. Meat can be cooked by a wide variety of methods including standard air-circulation ovens, rotary ovens and conveyor grills. The objectives of all methods are identical, i.e. to ensure all parts of the meat receive a heat process sufficient to achieve the correct organoleptic quality of product and to destroy contaminating vegetative microorganisms, in particular enteric pathogens such as *Salmonella* and pathogenic *E. coli*.

Meat entering an oven can vary significantly in size, density and temperature. In addition, the oven may vary in its heat distribution characteristics with hot and cool spots occurring inside. As a consequence, before supplying product for sale it is incumbent on processors of cooked meat products to undertake a full validation study of each different heat process and identify and implement appropriate controls at relevant critical points to ensure the heat process is applied consistently correctly.

It cannot be over-emphasised how important cooking process validations are to the ongoing safety of the process and product. Whilst it may be possible to generate safe product on occasion through visual checks and temperature probing of a number of products exiting the oven, this approach is as unreliable as end product microbiological testing for assuring the safety of the entire batch.

A properly conducted process validation need only be carried out once and the ongoing process controls, critical limits and associated process checks and required records can be established through this study. Periodic re-validation would be good practice but in any case, must, of course, be conducted if process or product parameters change, such as in the use of larger meat pieces, different meats, different containers, different temperature/time profiles or different ovens.

Process validation should establish the conditions necessary to achieve the desired, safe cooking procedure for the specific product under all worst-case conditions. Therefore, it must take into account, amongst other factors, all of the following:

- largest meat/piece size and density
- lowest ingoing meat temperature
- type of container/packaging
- coldest part of the oven
- maximum oven load
- shortest process time
- lowest process temperature

Temperatures in the oven chamber and in a number of products should be monitored using thermocouples attached to a continuous chart recorder. The product temperature should be measured in the slowest heating part of each product, which will usually be its centre. This should be repeated, preferably on two more occasions, i.e. three cooks, and the data then used to establish the critical limits and appropriate monitors of the critical parameters necessary to consistently achieve a safe cook. The following parameters would be expected to be included in the monitoring procedure for a cooked meat process:

- meat/piece size (a maximum size must be specified and controlled)
- ingoing temperature (coldest acceptable temperature must be specified and checked)
- load of the oven (a maximum load must be specified and complied with)
- position of product in the oven, if appropriate
- temperature of the oven (and product where possible) and length of time of the process (usually referred to as 'cooking profile')

Such factors must be monitored for each batch of product cooked using a representative sample. Checks should ideally be carried out using continuous on-line systems, such as check weighing of meat pieces and continuous temperature recording of raw meat refrigerators supplemented by regular temperature probing of product just prior to cooking. The process temperature and time of ovens should be monitored continuously using probes linked to a chart recorder and these systems should be fitted with a suitable alarm to warn if the process falls below the minimum temperature and time specified. All probes used to monitor or check the temperature of product during or after cooking should be routinely calibrated.

Where other factors are important for achieving the correct cook, e.g. the use of a steam supply, appropriate controls, monitors and alarms must be fitted to ensure cooking processes are not compromised by a steam failure or a fall in steam pressure.

It is common practice to measure end product temperature by manually probing some products immediately after the cook to give added assurance that all of the process controls employed during the cook have been effective. The minimum temperature specified at this point is usually a higher temperature than that critical to product safety to provide a margin of safety. Product not achieving this minimum temperature may be safe, but these 'caution' levels are useful as they can trigger an investigation to determine the reasons for not achieving the specified temperature and allow adjustments to be made before the process or product becomes unsafe. Because of the heat penetration characteristics associated with meat pieces of different sizes, it is often the smaller sized meat pieces that are prone to undercooking rather than the bulk meats. This is because the margin for error in temperatures and times in the former are far smaller than the latter. For example, whilst 70°C for two minutes may be set as the minimum process for both large/bulk and small size cooked meat products, once a bulk meat has reached 70°C throughout, because of the heat retention properties of large pieces of meat, it can remain at this temperature for long periods, far exceeding the two minutes required. However, a small piece of chicken cooked on a continuous conveyor under a grill is usually cooked for very short periods as it will reach temperature quickly. Once moved away from the heat source however, it will lose temperature rapidly and therefore the 'line' between safe and unsafe processing is much finer. Therefore, in general, small meat pieces are more at risk from undercooking than larger bulk meats.

Irrespective of the meat product type, the intention in the UK is to apply a cooking process equivalent to 70°C for two minutes, which is recommended as the minimum heat process required to achieve at least a 6 log reduction in contaminating vegetative pathogens such as *Listeria monocytogenes* and *Salmonella* (Anon, 1992a). In fact, this heat process achieves far in excess of a 6 log reduction in strains of *Salmonella*. In the USA, the minimum heat process required for cooking is 68°C for 15 seconds or an equivalent process for injected meats and 63°C for three minutes or equivalent for comminuted meat and fish products (Food and Drug Administration, 1999f). Poultry, wild game and stuffed meat, fish and poultry must be cooked to 74°C for 15 seconds and whole roast joints of meat must be cooked to achieve 63°C for three minutes or an equivalent process.

Studies conducted on the thermal destruction of *Salmonella* have shown that most serotypes are not especially heat resistant (Table 4.12). Murphy *et al.* (1999) inoculated a six-strain mixture of *Salmonella* serotypes into chicken and cooked the meat at different temperatures. They reported a *D*

Table 4.12 Reported *D* values of different *Salmonella* serotypes

Salmonella serotype	Temperature (°C)	*D* value (minutes)	*z* value (°C)	Medium	Reference
6-strain cocktail (Senftenberg, Typhimurium, Heidelberg, Mission, Montevideo, California)	67.5	0.286 +/− 0.001	NR	Ground chicken*	Murphy *et al.* (1999)
		0.261 +/− 0.002	NR	Peptone-agar†	
	70	0.176 +/− 0.005	NR	Ground chicken*	
		0.085 +/− 0.009	NR	Peptone-agar†	
5-strain cocktail (Newport, Typhimurium, Agona, Bovismorbificans, Muenchen, unknown)	51.6	61–62	5.56	Ground beef	Goodfellow and Brown (1978)
	57.2	3.8–4.2			
	62.7	0.6–0.7			

Table 4.12 *Continued*

Salmonella serotype	Temperature (°C)	*D* value (minutes)	*z* value (°C)	Medium	Reference
Senftenberg	53	53	6.25	Ground beef‡	Orta-Ramirez *et al.* (1997)
	58	15.17			
	63	2.08			
	68	0.22			
Senftenberg	55	227.18	5.4	Ground turkey§	Veeramuthu *et al.* (1998)
	60	13.55			
	65	3.08			

NR = not reported.

* 78% moisture, 0.12% fat (lipid), pH 5.9.
† 0.1% peptone, 0.1% agar, pH 5.7.
‡ 72.8% moisture, 3.8% fat, pH 6.0.
§ 74.4% moisture, 4.3% fat, pH 6.29.

value of 0.286 (+/− 0.001) minutes at 67.5°C and 0.176 (+/− 0.005) minutes at 70°C. The calculated D value at 70°C using a peptone-agar medium was 0.085 (+/− 0.009) minutes, about half that of the time required to achieve an equivalent reduction in the real food system. This phenomenon is a very common finding with food microorganisms; their behaviour in solid food being very different to that in liquid or non-food systems. The relatively high D values found in this study can be attributed to the most heat resistant strain used in the cocktail, *S.* Senftenberg. When inoculated into chicken breast meat, heated to 67.5°C (15 g sample in a 50 mm diameter by 8 mm height tube) and then immediately cooled, *S.* Senftenberg was reduced by 1.1 log unit, *S.* Typhimurium by 3.35 log units and the other four *Salmonella* serotypes by > 6 log units (Murphy *et al.*, 1999). Similar D values (D_{68} 0.22 minutes) were reported by Orta-Ramirez *et al.* (1997) for *S.* Senftenberg in ground beef (Table 4.12); they also reported a z value of 6.25°C. Based on this work, it is evident that the heat process advocated in the UK for cooking meats (70°C for two minutes) would give a reduction equivalent to *c.* 19 log units of *S.* Senftenberg (Table 4.13). As this organism is generally considered to be 5–10 times more heat resistant than most other *Salmonella* serotypes, the margin of safety in this process for most other *Salmonella* spp. is considerable. It should be noted that the increased heat resistance of *S.* Senftenberg in comparison to other *Salmonella* serotypes occurs only in foods of high water activity. As the water activity decreases, other *Salmonella* serotypes demonstrate equal or greater heat resistance than *S.* Senftenberg (see Chapter 3).

Applying the same calculation to the USA cooking requirements would give the following approximate reductions in *S.* Senftenberg: 68°C for 15 seconds (injected meats) would give 1.1 log reduction (Table 4.13), 74°C for 15 seconds (poultry, game and stuffed meat, fish and poultry) would give 10.4 log reduction and 63°C for three minutes (whole roast joints of meat and comminuted meat and fish products) would give 2.16 log reduction. In all cases, reductions in commonly occurring *Salmonella* serotypes would be in excess of 3 log units.

Doyle and Mazzotta (2000) have reviewed some studies on the heat resistance of *Salmonella* in food products and concluded that a process of 71°C would require 1.2 s to inactivate 1 log of *Salmonella* cells (z value 5.3°C). This estimate excludes *S.* Senftenberg and is very similar to the values calculated using the data of Goodfellow and Brown (1978) derived from work with a mixture of *Salmonella* serotypes (not *S.* Senftenberg).

Like most organisms, sensitivity to heat is dependent on the intrinsic

Table 4.13 Estimated log reduction of *Salmonella* for a variety of nationally recommended heat processes

Heat process	Products the heat process is applied to (country where guideline applies)	*Salmonella* serotype(s)	*D* value (*z* value (°C) used as reference)	Reference source (of *D* and *z* value)	Estimated reduction* (log cfu/g)
70°C, 2 min	Meat, fish, ready-meals (UK)	Newport, Muenchen, Bovismorbificans, Typhimurium, Agona, unknown serotype	$D_{62.7}$ 0.7 min (5.56)	Goodfellow and Brown (1978)	58.8
		Senftenberg	D_{68} 0.22 min (6.25)	Orta-Ramirez *et al.* (1997)	18.9
71.7°C, 15 s	Milk (UK)	Newport, Muenchen, Bovismorbificans, Typhimurium, Agona, unknown serotype	$D_{62.7}$ 0.7 min (5.56)	Goodfellow and Brown (1978)	14.8
		Senftenberg	D_{68} 0.22 min (6.25)	Orta-Ramirez *et al.* (1997)	4.4
68°C, 15 s	Injected meats (USA)	Newport, Muenchen, Bovismorbificans, Typhimurium, Agona, unknown serotype	$D_{62.7}$ 0.7 min (5.56)	Goodfellow and Brown (1978)	3.2
		Senftenberg	D_{68} 0.22 min (6.25)	Orta-Ramirez *et al.* (1997)	1.1

(Continued on p. 146.)

Table 4.13 *Continued*

Heat process	Products the heat process is applied to (country where guideline applies)	*Salmonella* serotype(s)	D value (z value (°C) used as reference)	Reference source (of D and z value)	Estimated reduction* (log cfu/g)
74°C, 15 s	Poultry, game & stuffed meat, fish & poultry (USA)	Newport, Muenchen, Bovismorbificans, Typhimurium, Agona, unknown serotypes	$D_{62.7}$ 0.7 min (5.56)	Goodfellow and Brown (1978)	38.5
		Senftenberg	D_{68} 0.22 min (6.25)	Orta-Ramirez et al. (1997)	10.4
63°C, 3 min	Comminuted meat and fish products and whole roast joints of meat (USA)	Newport, Muenchen, Bovismorbificans, Typhimurium, Agona, unknown serotype	$D_{62.7}$ 0.7 min (5.56)	Goodfellow and Brown (1978)	4.9
			D_{68} 0.22 min (6.25)	Orta-Ramirez et al. (1997)	2.2

*These are estimates only and do not take account of factors such as the heating medium used, the water activity or other factors such as heat shock response or 'tailing' which may affect the thermal destruction and survival of *Salmonella*.

factors of the product in which the organism is cooked. Juneja and Eblen (2000) demonstrated the potential protection afforded to *Salmonella* by the presence of fat in meat products. An eight-strain mixture of *Salmonella* Typhimurium DT104 inoculated into beef of different fat contents was heated to a defined temperature and the time to achieve a 7 log reduction was calculated. Taking account of the different lag times before reduction began (due to different heat transfer characteristics caused by the fat), beef with a fat content of 7% required 7.07 minutes at 65°C to achieve a 7 log reduction, whereas beef with 24% fat required 20.16 minutes to achieve the same reduction.

Low water activity products are also protective against thermal destruction of microorganisms and a low pH in products increases their sensitivity to heat (see Chapter 3). Goepfert *et al.* (1970) demonstrated this effect using eight different *Salmonella* serotypes, including *S.* Senftenberg (Table 4.14). The average *D* value at 57.2°C was 3.3–35 times longer at a water activity of 0.96 than at 0.99 when using sucrose as the humectant. At a water activity of 0.90, an equivalent reduction would require a heat process up to 75 times greater than at a water activity of 0.99. The protective effect was greater in the presence of sucrose, followed in order of effect by sorbitol and then fructose and glycerol.

It is evident from this work that heat processes applied to dry products must be extended (higher temperature or time or both of these) to achieve a similar level of safety to that achieved during normal heat processing of high water activity meat products. In such situations, it is

Table 4.14   Effect of different water activities on heat resistance of *Salmonella* serotypes at 57.2°C, adapted from Goepfert *et al.* (1970)

| | Mean $D_{57.2}$ value (minutes) | | | |
| *Salmonella* serotype | Water activity, a_w | | | |
	0.99*	0.96*	0.93*	0.90*
Infantis	0.9	5.3	14.3	21.1
Alachua	1.1	26.5	60.0	80.0
Typhimurium	1.1	14.3	30.5	46.7
Anatum	1.0	35.5	68.7	62.5
Anatum GF	1.1	NT	56.0	59.0
Montevideo	1.1	20.0	58.4	72.5
Senftenberg 775W	14.5	48.3	55.0	62.0
Tennessee	0.8	9.5	21.3	23.5

NT = not tested.
*Aqueous sucrose solution, pH 6.9.

recommended that challenge test studies are carried out in research facilities to determine the appropriate heat process necessary to achieve the desired reduction in pathogen levels.

This may also be necessary for any process that deviates significantly from normal practice. For example, the D value of *Salmonella* has been reported to increase if the organism is held at sublethal temperatures for extended periods (Bunning *et al.*, 1990; Xavier and Ingham, 1997). This is believed to be due to the production of heat shock proteins that give partial protection to the organism at high temperatures. A cooking process that is changed to include a slow heating-up phase may therefore be more vulnerable to the survival of the organism. In practice, an increase in the D value of c. 3-5 fold does not exceed the safety margin built into the processes currently used, but as safety margins are cut when minimal processes are developed, such factors may become increasingly important.

Mathematical models are available that can predict the survival of *Salmonella* under defined high temperature conditions. Such models are useful for assessing the general efficacy of different processes but they cannot take account of all factors that may affect the survival of the organism in a particular food. It is normal, when applying heat processes lower than the accepted industry norms to employ challenge tests, conducted in research facilities, to establish the efficacy of the proposed process against a large number of *Salmonella* serotypes.

Cooking is used to destroy any vegetative pathogens present in the raw materials. Having achieved this through adequate processes and effective process control, it is then essential to avoid re-contamination of the cooked meat product after the cooking stage and prior to packing. It is normal therefore for cooked meat manufacturers to completely segregate factories into two areas, one dedicated to handling raw meat and the other dedicated to handling the cooked product. To achieve this, the oven is usually sited in a separating wall that facilitates complete segregation of the factory into a raw and a cooked side. Ovens are 'double entry' so that the raw meat can be placed on racks and loaded into the oven from the raw side of the factory by operatives dedicated to raw meat preparation. Whilst the oven door on the raw side is open, it should not be possible to open the door on the cooked side and vice versa. The 'raw' door is closed and the product is cooked. After the cooking cycle is complete, the door on the cooked side of the factory, often referred to as 'high-risk' or 'high-care', is opened and the rack removed. Products are temperature probed and set aside at ambient temperature to allow the immediate heat to dissipate/'flash off' before they are handled and further processed by operatives and equipment dedicated to the high-risk side of the factory.

The high-risk side of the factory must be managed to prevent the entry of bacterial pathogens and, whilst this is hardest to achieve with organisms such as *Listeria*, it should be readily achievable with enteric pathogens such as *Salmonella*.

However, in order to achieve this, significant effort must be put into the design of the factory. This must include consideration of the drainage system, which should be designed to allow water to flow only from the cooked to the raw side of the factory and include appropriate water traps to prevent backflow of potentially contaminated water into the high-risk area. Air should be under positive pressure on the high-risk side to ensure it flows out of high-risk and into low-risk areas and that airborne contaminants do not enter the high-risk areas from any contaminated sources. In addition, strict personnel hygiene and control procedures must be operated to prevent people either bringing in pathogens from external sources or introducing such contaminants from themselves. Therefore, entry into the high-risk area must be via controlled changing rooms separated in the centre with a physical barrier such as a bench. A logical changing procedure should be maintained, e.g.

- remove shoes and outer clothing
- wash hands and put on appropriate hair covering
- step across the physical barrier into the high-risk side of the changing area
- put on shoes that are dedicated for use in the high-risk area
- wash hands
- put on a clean dedicated coat or outer garment
- enter the high-risk area
- re-wash hands and/or apply a bactericidal hand cleanser

These steps should be included as the minimum standards required for large-scale manufacturers of cooked meat products.

In addition to these simple procedures applied to personnel entering high-risk facilities, it is essential to ensure adequate procedures are in place to control infectious disease in operatives working with any ready-to-eat food, including cooked meats. Such control procedures are specified in guidance documents, e.g. Anon (1995a), and these include exclusion from work (for 48 hours after the first normal stool) if an individual is or has recently suffered from gastrointestinal illness involving vomiting or diarrhoea or has travelled to countries where they may be more susceptible to contracting an infection with specified agents. Under such circumstances it is prudent to employ stool testing for *Salmonella* before allowing vulnerable operatives to work in the high-risk side of the factory. There is generally little risk of workers who have suffered salmonellosis passing this on to other workers providing good standards of personal hygiene are reinforced and practised. However, they may present a higher risk to the cooked products and it may be considered prudent to employ them on the low-risk side of the factory until stool sample test results reveal absence of the organism.

Some manufacturers employ routine stool testing of operatives for *Salmonella* and do not allow those found to be carrying the organism to work with ready-to-eat foods. Whilst this certainly diminishes the risk associated with potential contamination of the food by personnel, it should be recognised that stool testing is not entirely reliable for detecting individuals carrying the organism asymptomatically as it can be shed intermittently. Indeed, such individuals do not necessarily present a hazard to the product, providing appropriate controls are in place in relation to personal hygiene practices. Clearly, policies relating to stool testing of individuals and associated action including exclusion from specific work activities must rest with individual food businesses.

One often-overlooked area of potential contamination in relation to personnel is that of toilet practices. Toilets clearly have the potential to harbour enteric pathogens and if inadequate changing and personal hygiene procedures are operated it may be possible to transfer contaminants from these areas into the factory. Toilets should be located in separate areas and staff should remove high-risk clothing including footwear which should be left in the high-risk changing room before they go into toilet areas. Clearly, the implementation of effective cleaning and sanitisation procedures in the toilet areas is also important to ensure these do not become a focus of contamination.

After initial heat dissipation from cooked products removed from the oven, they are placed in a forced-air (blast) chiller and cooled to $<5°C$. Some cooling systems used to dissipate the initial heat from the cooked product are more prone to introducing contaminants than others and particular attention should be given to water shower systems that are employed for pre-cooling purposes, usually for meat enclosed in cooking canisters. Meat debris can accumulate in the bottom of these water-cooling systems and contamination can build up as the water is recirculated. These water-cooling systems should be regularly cleaned and disinfected and low levels of chlorine should be maintained in the water to ensure the water does not become a source of contaminants.

Following 'flash' heat dissipation, the time taken to bring the product to chill temperatures varies according to its size but this is of concern in relation to spore-forming bacteria and is not an important consideration in relation to the control of *Salmonella*, as the organism should not be present in the first instance having been destroyed by the cooking process. Large bulk meats cooked in bags or containers are usually knocked out of their containers or removed from their bags soon after cooking and prior to blast chilling. In some cases they may be blast chilled first and then removed from their containers. Other products such as whole chicken to be used for pieces or portions, are cooked whole and then either stripped of meat whilst still warm or blast chilled and then stripped or portioned as required.

A number of products, after removal from their containers or bags and cooling are subjected to a further short time, high temperature roasting stage. The intention is to impart a roasted appearance to the outside of the meat and, whilst the oven or flame tunnel reaches temperatures in excess of 200°C, the surface of the product reaches pasteurisation temperatures (> 70°C) and the inside only achieves temperatures approaching 30°C. This stage is useful for destroying any vegetative contaminants present on the surface of the meat but introduces further hazards associated with the potential for growth of surviving spore formers that may be present in the bulk of the meat. After roasting, these products are again blast chilled to < 5°C.

Clearly, there are many opportunities for microbial contaminants to gain access to products during post-cooking processes. The effective implementation of appropriate hygienic practices and cleaning procedures for equipment is essential to the safety of these products. Handling of products and product contact surfaces by personnel represents one of the main routes by which enteric pathogens can be transferred to the product and this must be consistently well controlled. In addition, microbial contaminants can enter and build up on inadequately cleaned and maintained equipment and attention to detail during maintenance, cleaning and sanitisation is absolutely essential. Equipment that is particularly difficult to clean includes the feeder chute to the meat slicer and the slicing unit itself. In addition to these, other areas or items of equipment that have significant potential to introduce contaminants include chiller units, particularly where the product is exposed, knocking out/debagging tables, and the slicer conveyor. All of these areas must receive particular and thorough attention during cleaning. Other areas such as floors, walls, drains, cleaning equipment and items coming into contact with hands such as door handles and cleaning utensils should not be overlooked during cleaning operations. Whilst *Salmonella* is unlikely to be a contaminant in the high-risk area, if it is introduced into the environment, the effective cleaning and sanitisation of all the above-mentioned areas is absolutely essential to ensure its adequate removal.

In addition to the potential routes of microbial contamination described, another potential source is the waste handling area of the high-risk environment. Careful consideration should be given to the control of waste material exiting the area. Chutes are a useful means of sending waste material out of the area although the chute must be regularly cleaned and sanitised to prevent this from becoming a source of contamination. It is common for waste bins or trolleys to be used in high-risk areas and if removed from the area for purposes of emptying, appropriate procedures for cleaning and sanitising these containers for return to the high-risk area must be in place. Guidance on appropriate procedures for the operation of high-risk manufacturing environments are available (Chilled Food Association, 1997 and 2001).

Some cooked meat manufacturers, because of their small size, do not have the types of segregation and procedures described above. It is also common for butchers' shops to handle both raw and cooked meats; sometimes the cooked meat product is made in the same facility as the raw meat butchery. These shop units are small and the number of available staff does not allow for separate individuals to be dedicated to raw or cooked meat handling.

Clearly, where raw and cooked meats are handled and cooked in the same areas by the same personnel, the risk of cross-contamination is significantly increased. Nevertheless, it should still be possible to minimise cross-contamination by adopting similar principles to those described above. The principles of keeping raw foods separated from cooked foods by using different utensils and surfaces for each and the application of strict personnel hygiene procedures and practices between handling raw and cooked foods will minimise the chances of cross-contamination. If raw and cooked meats are handled in the same room, they should be prepared at different times and all surfaces should be fully cleaned and sanitised between the two operations. Clearly, it is preferable and safer to prepare raw and cooked meats in separate rooms with proper control of product and personnel between the rooms. Failure to operate such principles effectively have been identified as a significant contributory factor in food poisoning outbreaks and was a significant factor in the fatal *E. coli* O157 outbreak in Lanark, Scotland (Ahmed, 1997).

Final product issues

Cooked meat products are sold as bulk units for slicing on the delicatessen counter or as pre-pack units for retail display. The majority of bulk packs are sealed under vacuum whereas many pre-packs are packed under a modified atmosphere with low residual oxygen. Indeed, it is becoming common for oxygen scavengers to be placed inside the pack to help achieve longer product shelf life.

Prior to packing, some products are garnished with herbs or spices and occasionally some fruit, as in the case of bulk packs of pâté. Such materials must be under tight control in respect of enteric pathogens, as many herbs and spices are grown and harvested under conditions in which enteric pathogens such as *Salmonella* could gain entry. Therefore, any items of garnish taken into the high-risk area of the factory have the potential to introduce microbial contaminants into that area. Ideally, such materials should be previously heat treated to destroy vegetative contaminants, but in many cases this is not possible for organoleptic reasons. To further help reduce contamination entering the high-risk area, some garnish materials, either inside or unpacked from their containers as appropriate, may be passed through a bath of a chlorinated water (> 50 ppm free chlorine) into the high-risk area.

In addition to herbs and spices, pâté is often coated with molten gelatine to seal the surface. This too can be a potential source of enteric pathogens and should be pasteurised or suitably controlled to ensure such contaminants are not present. The conditions of use of gelatine need careful control as the tanks are often kept under warm conditions for many hours of continuous use. Tanks should be emptied and cleaned regularly. The use of more ambient stable varieties of gelatine, e.g. pH controlled, can contribute to the control of contaminants in this vulnerable area.

The final formulation of the cooked meat is of little relevance with respect to *Salmonella* as the organism should not be present. In general, cooked meats have a neutral pH varying from *c.* 6.0 for pork and beef to *c.* 7.0 for poultry products. The aqueous salt varies significantly from 1% for poultry products to in excess of 2.5% for many cured hams. The shelf life of these products varies significantly depending on the type of product. Shelf lives are allocated with due consideration given to psychrotrophic pathogens such as *Listeria* and *C. botulinum* (non-proteolytic types) together with anticipated spoilage microflora. Cooked, uncured meat such as poultry and beef may have chilled shelf lives of < 10 days to 20 days. Cured meats such as pâté and ham may have shelf lives ranging from 15 to 30 days or more. Some canned, cured, perishable meat products are given shelf lives of several months. However, none of the shelf lives are set with *Salmonella* as a consideration as this organism should not be present. Although there are reports of growth of *Salmonella* serotypes at temperatures of 7°C or less, under effective chill temperature control (< 8°C), the organism, if present, would not be expected to increase significantly if at all during the shelf life of these products.

Although pre-packed meat is not subject to any further contamination after leaving the factory, bulk meats to be sliced on the delicatessen counter may be exposed to a significant number of additional potential sources of contamination. All the efforts of the manufacturer of the cooked meat product to destroy contaminants and avoid re-contamination can be easily undone by poor hygienic practices in the meat display area of the delicatessen counter. The potential for cross-contamination from raw meat or other raw foods must be effectively controlled. Raw foods, especially raw meat should not be displayed next to cooked meat and effective segregation between them must be in place. Separate utensils and weighing equipment must be used for each and personnel handling practices must be under very strict control especially if they have to handle both raw and cooked meats. Ideally, separate staff should be employed to handle raw and cooked products. A dedicated hand-wash basin suitably supplied with hot water, soap and disposable towels should be readily available for staff to use. Cleaning and sanitisation of delicatessen counters and display containers and utensils must be carried out regularly and effectively to avoid the retail environment becoming the point at which microbial contamination accesses the product.

Staff training is crucial if cooked meats are to be supplied consistently safely. In both the manufacturing and retail environments, staff handling food should be fully trained and kept up-to-date in all appropriate food safety procedures and personal hygiene practices.

Many of the good practices discussed herein were recommended as appropriate measures for the prevention of further outbreaks following an investigation into the extensive *E. coli* O157 outbreak in Lanark, Scotland (Pennington, 1997). The introduction of licensing of meat butchers in the UK in 2000, for which training of staff in food hygiene and segregation of raw and cooked foods are required before a licence is granted, are welcome steps towards food safety assurance. Whilst many responsible food retail outlets have always operated such principles, the sad fact is that legislation has been required to enforce the operation of such obvious and simple controls in all such food outlets.

A number of surveys for the presence of enteric pathogens in cooked meat products have been reported. In a survey of cooked meat products in the UK (Anon, 1996c), *Salmonella* was not detected in any of 414 samples (Table 4.15). A similar survey of cooked, chilled chicken products on retail sale in England and Wales (Anon, 2000f) found *Salmonella* in two (the drumstick on two whole chickens) of 758 samples (0.3%). Surveys of cooked meats have also been carried out in other countries. In a survey of cooked foods in Singapore, Ng *et al.* (1999) reported the detection of *Salmonella* in 24 out of 1070 (2.2%) samples of cooked meat, cooked fish and cooked meat containing products. *Salmonella* was not detected in 14 samples of cooked hamburgers sampled as part of this survey.

Microbiological surveys do occasionally detect *Salmonella* in cooked foods including cooked meat products (Table 4.16) and, in this day and age, it is unacceptable for cooked, ready-to-eat foods to be presented to the consumer containing viable enteric pathogens. The results of a survey of cooked meat manufacturing practices reported in 1995 (Anon, 1995b) clearly support the need for adequate training of personnel involved in the manufacture of cooked meat products and serve to underline some of the factors that may contribute to the occasional presence of *Salmonella* found in microbiological surveys of cooked meat products. A third of the processors surveyed were not able to state the temperature reached at the centre of their product during cooking and 36% of processors did not calibrate their temperature probes, so it is not surprising that some failures in cooking may occur. In the same survey, 27% of processors were reported to handle raw and cooked foods in the same area and although the procedures in place to prevent cross-contamination in these areas were not reported, it is clear that inadequacies in segregation between raw and cooked product or associated personnel, would readily lead to enteric pathogens passing from the raw to the cooked product.

The safety of cooked meat products in relation to enteric pathogens such as *Salmonella* should be readily achieved through the implementation of some fairly simple systems and procedures involving raw material quality assurance, application of effective cooking processes and reliable segregation of raw from cooked product (including equipment, utensil and personnel control and personal and environmental hygiene practices). Application of a simple hazard analysis would identify such areas as critical to the safety of these products and, whether a small or large manufacturer

Table 4.15 Survey of the incidence of enteric pathogens and indicator organisms in cooked, sliced, pre-packaged meat products, adapted from Anon (1996c)

Product type	Enterobacteriaceae (log cfu per g) (number of samples in each count range)									Salmonella/25 g
	<1	1–1.99	2–2.99	3–3.99	4–4.99	5–5.99	6–6.99	7–7.99	8–8.99	
Cooked, cured pork	135	22	21	16	13	7	4	1	1	0
Cooked, cured, comminuted meat	34	3	5	2	1	1	0	1	0	0
Cooked poultry	42	4	10	7	8	11	13	4	0	0
Corned beef	16	5	5	9	5	4	4	0	0	0

Table 4.16   Surveys of the incidence of *Salmonella* in cooked meat products

Product	Number of samples	*Salmonella* detected in 25 g (no. of samples)	Percentage positive (%)	Reference
Cooked, cured pork	220	0	0	Anon (1996c)
Cooked, cured, comminuted meat	47	0	0	
Cooked poultry	99	0	0	
Corned beef	48	0	0	
Cooked chicken	758	2	0.3	Anon (2000f)
Cooked meat and meat products (beef, mutton, pork and chicken)	403	7	1.7	Ng *et al.* (1999)
Meat or vegetable containing cooked rice	223	5	2.2	
Cooked seafood	430	12	2.8	
Cooked hamburgers	14	0	0	

or retailer, everyone should be capable of applying the controls necessary to avoid exposing consumers to pathogens such as *Salmonella* in these finished products.

CHOCOLATE

Chocolate products are probably the most widely consumed of confectionery products in the world. They have been implicated in a number of outbreaks of salmonellosis in the last few decades (Craven *et al.*, 1975; D'Aoust *et al.*, 1975; Gill *et al.*, 1983) and are one of the food types in which the infectious dose of the organism has been demonstrated to be exceptionally low. Chocolate is the product of cocoa beans and is an almost inert finished product containing little moisture. It is consumed by all sectors of society and is cherished particularly by children – and most adults! It is sold as chocolate bars, eggs, coins and a huge array of associated novelty products in isolation or in combination with other confectionery ingredients, e.g. toffee, biscuit, flavoured fondants and jellies, liqueurs etc., for which it is most frequently used as a coating, e.g. chocolate-coated bars.

Description of process

Chocolate is manufactured from cocoa beans. Cocoa beans are harvested, fermented, dried and then transported to a processing facility. After receipt (Figure 4.4), the beans are cleaned to remove foreign objects and insects and then roasted. The roasting achieves temperatures in excess of 110°C. The beans are air-cooled and then the seed coat is removed from the seed (nib). The nibs are ground to form chocolate liquor, which is held molten at temperatures between 60°C and 80°C. The liquor may be mixed with other ingredients to form different cocoa products. For example, it can be mixed with sugar and dried milk to form chocolate crumb or pressed to form cocoa press cake and cocoa butter. In order to make chocolate, the warm liquor mixture is held for extended periods, up to 18 hours and gently mixed continuously in a process known as conching. Chocolate is then held in warm storage tanks at *c.* 40°C and fed into moulds and cooled to form chocolate bars or a variety of different shaped products. In many cases it will be formed into large bars for distribution to other confectionery manufacturers where it may be mixed with other ingredients and used as a coating for fondant-filled chocolates or chocolate-coated confectionery bars. Chocolate products are usually wrapped in paper, plastic or foil or packed in boxes which are sealed and are stored and sold at ambient temperature.

Chocolate has an exceptionally low moisture content of <8% and consequently a very low water activity (0.3–0.5). As microorganisms do not grow at water activity values below *c.* 0.60, chocolate does not support microbial growth. It is allocated a long shelf life of several months that is

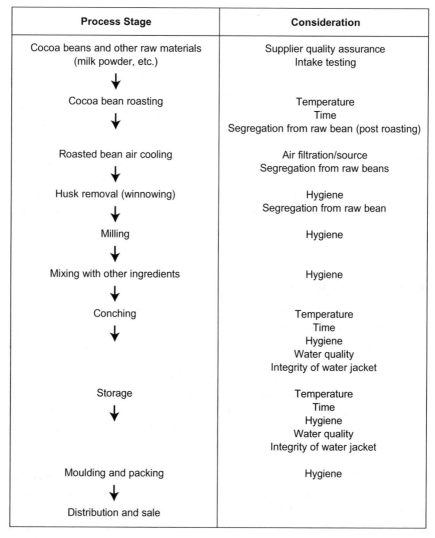

Process Stage	Consideration
Cocoa beans and other raw materials (milk powder, etc.) ↓	Supplier quality assurance Intake testing
Cocoa bean roasting ↓	Temperature Time Segregation from raw bean (post roasting)
Roasted bean air cooling ↓	Air filtration/source Segregation from raw beans
Husk removal (winnowing) ↓	Hygiene Segregation from raw bean
Milling ↓	Hygiene
Mixing with other ingredients ↓	Hygiene
Conching ↓	Temperature Time Hygiene Water quality Integrity of water jacket
Storage ↓	Temperature Time Hygiene Water quality Integrity of water jacket
Moulding and packing ↓	Hygiene
Distribution and sale	

Figure 4.4 Process flow diagram and technical considerations for a typical chocolate product.

only restricted by organoleptic deterioration rather than any microbiological reason. Although microbial growth is not supported, contaminating microorganisms can survive for very long periods of time in chocolate. Microbial growth can occur if local conditions of moisture and water activity increase due to poor storage conditions or if high water activity fillings are used for fondant-filled chocolates.

Raw material issues and control

The main raw ingredient used in the manufacture of chocolate is cocoa beans. Cocoa beans are widely recognised as being a potential hazard in relation to the introduction of *Salmonella* into chocolate manufacturing plants. Cocoa beans are grown in a variety of countries, particularly in South America and Africa and, in some areas, are grown and harvested under poor hygienic conditions. The beans are harvested from pods and together with the pulp surrounding them, are fermented in large mounds or containers. The pulp contains approximately 10% fermentable sugar and 80-90% water whereas the cocoa bean comprises about one third water, one third fat (cocoa butter) and the remainder sugars, starch, flavour compounds and other trace components (International Commission on Microbiological Specifications for Foods, 1998). The microorganisms contributing to the fermentation comprise a wide mixture of types and arise from the soil, air, the seed pod and vegetation, personnel and equipment. Yeasts that rapidly ferment the sugars to ethanol dominate the early stages of fermentation. Lactic acid bacteria, including streptococci, pediococci and lactobacilli are able to utilise the sugars to produce organic acids, e.g. lactic acid, and consequently increase in number. Acetic acid bacteria, capable of utilising ethanol are also to be found in high numbers together with a variety of moulds. The metabolic activity of microbial fermentative reactions generates significant heat within the mass causing the temperature to increase to in excess of 50°C. By the end of the process which lasts several days, the pulp is reported to increase in pH from *c.* 3.7 to pH 5-7, whereas the cocoa bean pH decreases from pH *c.* 6.0-7.0 to pH *c.* 5.0. Levels of *Bacillus* spp. increase in the pulp as the pH rises. The fermentation process and associated enzymatic changes achieve a number of important changes in the cocoa bean, particularly in relation to enhancing flavour, but it also destroys the germination capability of the seed.

After fermentation, the beans are dried. This is most commonly done in the sun; this process taking several weeks or, in some cases, they may be initially dried in the sun and then further dried in drying chambers. Drying is often conducted in areas subject to little environmental control and microbial contaminants can readily gain access to the drying bean. The beans are dried to a moisture level of less than 8% to avoid growth of mould during subsequent storage and transport.

Raw cocoa beans have a high microbial loading, mostly derived from the fermentation and drying process. *Bacillus* spp. represent the highest proportion of contaminants (*c.* 90%) and most other microbial groups are also represented including Enterobacteriaceae, lactobacilli, yeasts and moulds. Few surveys have been published of the incidence of *Salmonella* in raw beans but it is clear from salmonellosis outbreak investigations that raw beans are a source of the organism, although the frequency of contamination is likely to vary significantly. This will depend mostly on the hygienic conditions employed during the harvesting, fermentation and drying of the beans. The application of high standards of agricultural and process control in the primary processing of cocoa beans will reduce the risk of *Salmonella* being present in the raw bean and thereby reduce the risk of it entering the chocolate-making process and subsequently contributing to foodborne salmonellosis outbreaks.

Raw beans should be routinely screened for indicator organisms such as *E. coli* and specifically for *Salmonella*. This should be done to ensure that beans of consistently poor quality are not being received and the results can serve as a useful focus for discussion of microbiological quality between the supplier and the purchaser.

A variety of other ingredients are also used in the manufacture of chocolate and chocolate products, some of which are known to be potential sources of *Salmonella*. Milk concentrate and powder are frequently used in the production of milk chocolate. Milk powder has itself been implicated in a number of large salmonellosis outbreaks and the potential for this material to harbour *Salmonella* must be considered as part of the hazard analysis process. Sourcing from reputable manufacturers applying HACCP-based systems in the production of the milk powder and concentrate should be an absolute prerequisite for supply and the material itself should be subject to a significant level of microbiological testing for *Salmonella* and indicators of contamination such as Enterobacteriaceae or *E. coli* by the supplier. These requirements should be supported by regular inspection of the supplying factory as part of a raw material quality assurance scheme and routine intake (point of receipt) testing prior to use. Other ingredients used in chocolate manufacture such as sugar are not known to be sources of contamination of this organism.

Although not principally within the scope of this topic, a variety of filled and coated products are made using chocolate as the outer coating. Many of the fillings used for these products such as nuts, dried fruit, coconut and natural colours and flavours may be more important potential sources of *Salmonella* than the chocolate itself and should be controlled through an effective raw material quality assurance programme.

Once received at the factory, raw cocoa beans must be treated as a potential source of *Salmonella* and should be segregated from other materials that are to be used without any further significant *Salmonella* reduction process being applied, e.g. milk powder, sugar, etc. It is usual to store cocoa beans in silos prior to their being cleaned and further processed. Procedures and facilities allowing effective segregation of the raw beans from all further processed beans, bean products and ready-to-eat ingredients should include consideration of personnel, equipment, containers and air sources coming in contact with these.

Process issues and control

The first stage of the chocolate manufacturing process is to remove foreign objects and insects and physically clean the cocoa beans. This is done using a combination of magnets, air blowing and mesh screening. After this stage the beans are transferred into roasting chambers where they are heated using dry heat to temperatures above *c*. 110°C and up to 150°C. The beans are usually roasted for time periods ranging from 30 minutes to two hours; during this stage vegetative microorganisms can be reduced in number. At the same time the moisture of the already dry bean reduces even further to levels below 3% and the flavour of the bean is also enhanced.

Roasting is one of the most critical stages in the chocolate-making process as it is the principle stage where reduction in microbial contaminants can take place. It should be remembered that as the cocoa bean is already at a very low moisture level the roasting stage only achieves a limited reduction in contaminating organisms. Clearly, this reduction will vary significantly depending on the actual roasting temperatures and times employed. Roasting cannot be carried out at too high a temperature because of the adverse effect this has on flavour.

Barrile *et al.* (1971) studied the survival of general microflora on beans before and after roasting for 30 minutes at various temperatures. They found that levels were reduced by *c*. 1 log unit at 135°C, by *c*. 2 log units at 150°C and no microorganisms were detected after roasting at 180°C (Table 4.17).

It is anticipated that levels of *Salmonella* would decrease by a similar or perhaps, greater extent. However, it is clear that high levels of microbial contamination in the raw bean could exceed the capacity of the roasting stage to render the material free from contamination and therefore raw material quality remains of utmost importance. The use of steam heat in the roasting process has the potential to significantly increase the

Table 4.17 Survival of microorganisms on cocoa beans after heating at various temperatures for 30 minutes, adapted from Barrile *et al.* (1971)

Temperature of roasting (°C)	Number of microorganisms (cfu/g)*
Unroasted	1.5×10^7
135	2.6×10^6
150	2.7×10^5
165	2.8×10^3
180	0

* Mean of three trials.

destruction of *Salmonella* on the external surface of the cocoa bean due to the greater efficacy of wet heat on the destruction of microorganisms. Short bursts of steam into the beans at the roasting stage have been reported to give an additional 2–3 log reduction in the level of general microflora on the bean (D'Aoust, 1977). As the beans are subject to high temperatures and subsequent air-cooling, it may be possible to enhance the destructive effect of the heat process on microorganisms using some steam heat whilst maintaining the required low moisture content in the bean in the air-cooling stage although the introduction of steam may affect the roast quality of the bean. An alternative approach is to apply steam heat to decontaminate the nib just after the shell has been removed.

After roasting, beans are usually cooled with forced air. The air used for such purposes must not be sourced from areas common to the raw bean receipt and handling and should ideally be filtered. It would clearly defeat one of the achievements of roasting, if contaminants reduced during the roasting stage were to be re-introduced during the air-cooling stage through poor factory design and operation. Once cooled, the beans have their shells removed in a process known as winnowing. Like air-cooling of the beans, this process should be conducted in areas separated from those used for storing and handling raw beans.

Indeed, all areas after the roasting stage should be completely segregated from the raw bean areas. All airflow, personnel and equipment should be separated into the two areas, raw bean and post-roasting, and movement between the two should be subject to strict controls. This should include outer clothing and footwear changes for personnel, full height dividing walls and separate air supplies.

Outbreaks of salmonellosis have been attributed to chocolate becoming contaminated after roasting due to the roasted beans being further processed in the same room as the raw beans and air used for cooling also being drawn from the environment in which raw beans were present (Craven *et al.*, 1975).

After separating the shell from the cotyledons (nibs), the nibs are ground and milled to form chocolate liquor. This can then be pressed to form cocoa press cake and cocoa butter or mixed with sugar and dried milk to form chocolate crumb. Chocolate liquor is held in a molten state between 60°C and 80°C, mixed with other ingredients if required, and then passed through roller refiners and into conches. The conching process, which consists of slow continuous stirring of the chocolate to develop the correct consistency, lasts for between 5 and 96 hours at temperatures between 50°C and 70°C; it then passes to heated storage tanks (*c.* 40°C) prior to being moulded into bars or shapes or used to coat other fillings.

Chocolate is held for long periods of time at temperatures that would normally be expected to destroy contaminating vegetative pathogens such as *Salmonella*. However, after grinding and milling, chocolate liquor has a moisture content below 3% and after conching this can decrease even further to 1–1.5%. At such low water activities (0.3–0.5), the effect of the temperatures used in the process on microorganisms is almost negligible. Vegetative contaminants such as *Salmonella* are afforded almost complete protection against the destruction processes that readily occur at the same temperatures in aqueous based systems. With little or no water present in chocolate, the heat process has little effect on the organism. The protective effect afforded by chocolate has been shown in studies of the heat resistance of *Salmonella*.

Studies conducted by Goepfert and Biggie (1968) demonstrated that in molten chocolate *S.* Typhimurium had a *D* value at 70°C of 720–1050 minutes (12–17.5 hours) (Table 4.18), i.e. in order to achieve a 1 log unit reduction, it would take 12–17.5 hours at 70°C in this chocolate. Studies by Lee *et al.* (1989) confirmed the heat resistance of several *Salmonella* strains in milk chocolate with reported *D* values at 71°C of 4.5, 4.6 and 6.6 hours for *S.* Eastbourne, *S.* Senftenberg and *S.* Typhimurium, respectively (Table 4.19).

Therefore, whilst conching clearly could achieve a slight reduction in initial levels of microbial contaminants, the reduction is very small and cannot be relied upon to decontaminate the chocolate. However, together with the roasting stage, the combined reduction is useful to the overall safety of the product.

Table 4.18 *D* values of *S.* Typhimurium and *S.* Senftenberg in chocolate, adapted from Goepfert and Biggie (1968)

Temperature (°C)	*D* value (minutes)	
	S. Typhimurium*	*S.* Senftenberg†
70	720–1050	360–480
80	222	96–144
90	72–78	30–42

* z value 19°C.
† z value 18°C.

Barrile *et al.* (1970) studied the thermal destruction of *Salmonella* following the addition of small amounts of moisture. They demonstrated that milk chocolate inoculated with *Salmonella* (100 cells/100 g) could be effectively decontaminated (0 cells/100 g) at temperatures as low as 71°C after 10 hours if sufficient water was added to the chocolate to raise the moisture content to 3.7% at the start of heat treatment. They found that 2% moisture, added to the chocolate to raise the initial moisture content to 3.7% would achieve a 90% (1 log) reduction in *S.* Anatum within four hours, compared to 20 hours at the original moisture level. The added water evaporates during the process and results in chocolate of the same initial moisture content. The effect of different heating times at 71°C on *Salmonella* in milk chocolate with different initial levels of contamination is shown in Table 4.20.

Table 4.19 Reduction of *Salmonella* species in chocolate held at 71°C, adapted from Lee *et al.* (1989)

Time (hours)	Reduction (log cfu)		
	S. Eastbourne	*S.* Typhimurium	*S.* Senftenberg
0	0	0	0
2	0.5	0.3	0.5
4	0.8	0.7	0.6
6	1.4	0.9	1.3
8	1.7	1.2	1.6
10	2.2	1.6	2.2
12	2.6	1.8	>4.0
24	>4.6	2.7	>4.0
D_{71} value (hours)	4.5	6.6	4.6

Table 4.20 Survival of *S.* Anatum in milk chocolate with 2% added water at 71°C, adapted from Barrile *et al.* (1970)

Original inoculum level (MPN/100 g)	Heating time (hours)	Level of *S.* Anatum (MPN/100 g) after heating
3.4	2	2.1
	4	0.71
	6	0
29	4	1.1
	6	0.36
	8	0
100	6	1.4
	8	0.62
	10	0

After conching, chocolate is held molten under warm conditions (*c.* 40°C) until required for moulding. Warm water is used extensively in chocolate manufacture in the jacketing of holding vessels and pipes to keep the chocolate in a molten state. Poorly maintained pipes and vessels could allow water and associated microorganisms to contaminate the chocolate. Water used for such purposes should be of potable water quality and should be routinely disinfected and monitored for indicators of contamination using microbiological tests, e.g. for the presence of coliforms. In addition, the integrity of vessels and pipes must be routinely checked to ensure that there is no opportunity for leakage of water from these into the chocolate. Clearly, such a failure is more likely to have an adverse effect on product quality and be identified in this way, than to introduce *Salmonella,* but the potential for such contaminants to be introduced by this means should not be overlooked.

It is clear that once *Salmonella* is introduced into the chocolate process it will be extremely difficult to control and to eliminate. Chocolate manufacturers do not routinely employ wet cleaning systems as the entire process is managed with the aim of keeping moisture out of the product. Cleaning usually involves dry cleaning methods such as manually scraping down equipment and vacuum cleaning systems. In the event that wet cleaning is employed, it is absolutely essential to ensure that all cleaned and sanitised chocolate contact surfaces are fully dry before restarting production. Hot, molten cocoa butter, flushed through the system has been used to 'clean' equipment as part of a regime to minimise the use of water-based cleaning procedures.

It is common practice for chocolate manufacturers to 're-work' chocolate (product that is mis-shapen, broken, etc.) back into the process by adding small quantities of product collected from the end of the production line back into the mixing tanks. It should be recognised that certain risks are inherent with this practice as it continuously recycles material together with associated contaminants back through the system. If 're-work' must be used it should be routinely tested to ensure that microbial contaminants are not building up. If the microbiological results from this material indicate poorer than normal quality then 're-work' should be temporarily discarded to allow fresh 'virgin' product to pass through the system. In this way, the risks to safety associated with this practice can be minimised.

Final product issues

After moulding and setting, the solid chocolate product is wrapped in foil, paper or plastic and/or placed in trays in a box which is then sealed. Chocolate, with such a low water activity (0.3–0.5), is ambient stable and is allocated a shelf life of several weeks or months. Shelf life is limited by organoleptic deterioration, which occurs in both the taste and appearance of the product over time.

Salmonella, if introduced at any stage in the process after roasting from raw beans via other raw ingredients or through poor operator hygiene, will survive in the chocolate for extremely long periods of time.

Barrile *et al.* (1970) studied the survival of *Salmonella* in chocolate. When lyophilised cells of *S.* Anatum were inoculated into milk chocolate at a level of *c.* 50 cells per 100 g, the organism was still detectable at levels of 14.1 MPN/100 g after the 15-month storage period at room temperature (Table 4.21).

Table 4.21 Survival of *S.* Anatum, inoculated into milk chocolate at *c.* 50 cells per 100 g, over a 15-month period at room temperature, adapted from Barrile *et al.* (1970)

Storage time (months)	MPN *S.* Anatum (per 100 g milk chocolate)
5	54.2
8	34.5
12	14.1
15	14.1

Tamminga *et al.* (1976) studied the survival of *S.* Typhimurium and *S.* Eastbourne in artificially contaminated milk and bitter chocolate. Table 4.22 summarises their results. The combined evidence of these studies clearly indicates that *Salmonella* will survive for months in different types of chocolate.

Table 4.22 Survival of *Salmonella* in milk and bitter chocolate over a 9-month storage period at 20°C, adapted from Tamminga *et al.* (1976)

	Level of *Salmonella* (log MPN/100 g)			
	S. Typhimurium		*S.* Eastbourne	
Storage period	Milk chocolate*	Bitter chocolate†	Milk chocolate‡	Bitter chocolate§
0 days	5.04	4.86	5.2	5.2
1 day	2.34–2.63	1.69–1.88	4.64	4.64
13 days	1.18–1.36	0.30–0.56	2.54–3.18	1.30–1.90
20 days	0.89–1.11	Neg–0.30	2.54–2.97	1.18–1.56
34 days	Neg–0.89	Neg	ND	ND
41 days	ND	ND	2.23–2.38	0.65–1.18
48 days	Neg–0.89	Neg	ND	ND
76 days	ND	ND	1.63–1.69	Neg–1.46
83 days	Neg–0.30	Neg	ND	ND
6 months	Neg	Neg	Neg–1.23	Neg
9 months	ND	ND	0.89–1.11	Neg

Neg = *Salmonella* not detected.
ND = not determined.
* $a_w = 0.37$.
† $a_w = 0.42$.
‡ $a_w = 0.38$.
§ $a_w = 0.44$.

Following an outbreak of salmonellosis associated with imported chocolate-coated confectionery bars in the UK caused by *S.* Napoli, the outbreak strain, which was present at levels of 2–23 organisms per gram, was still detectable after 12 months (Gill *et al.*, 1983). Indeed, it is reported that many of the infections were contracted following consumption of these products over seven months after they were manufactured.

Therefore, whilst *Salmonella* will clearly not grow in chocolate, it can survive for extremely long periods of time. The introduction of even low levels of contamination is a significant hazard that must be prevented.

It is usual to subject cocoa and chocolate to high levels of testing for indicators of microbial contamination and also for the presence of

Salmonella. In addition to finished products, testing should be extended to raw materials including raw beans, roasted beans and environmental samples. Results from the microbiological examination of floor sweepings, spillages, scrap chocolate and chocolate 're-work' can serve as very useful indicators of the presence of and potential for *Salmonella* to gain access to the finished product. Testing cannot give assurance of freedom from contamination but, if used properly, it can help identify sources of contamination before they become out of control. It should also be remembered that *Salmonella* isolation from chocolate can be difficult and inhibitory compounds present in the chocolate need to be 'neutralised' using appropriate enrichment media (see Chapter 6).

Another important factor that increases the risk of salmonellosis associated with chocolate products is the apparent low infectious dose. It is evident from a number of outbreaks that very low levels of *Salmonella* present in chocolate are capable of causing illness. In the UK outbreak associated with chocolate-coated confectionery bars, the level of contamination ranged from 2 to > 23 organisms per gram of chocolate (Gill *et al.*, 1983). As only 3 g of chocolate coating were present on the confectionery bar, the levels capable of causing illness were in the region of 6 to > 66 *S.* Napoli. In an outbreak in the USA and Canada caused by *S.* Eastbourne, levels of *Salmonella* were estimated to be present at 2-9 *S.* Eastbourne per contaminated chocolate ball and the infectious dose was believed to be less than 100 cells (D'Aoust *et al.*, 1975).

The very low infectious dose associated with chocolate is believed to be due to the protective effect afforded to the organism during its passage through the stomach. The organism, which is probably encapsulated in fat, is protected from stomach acids and therefore the low numbers of cells in the chocolate can pass into the intestine where they cause an infection.

The other important consideration in relation to the increased risk of infection from contaminated chocolate is its consumption by vulnerable groups. Whilst all groups of society enjoy the pleasure of consuming chocolate, there is no doubt that it is particularly cherished by young children who are clearly a highly vulnerable group in respect of infectious intestinal agents.

The safety of chocolate and chocolate products is dependent on a number of factors including raw material control, process segregation and control (in particular of the roasting process), control of water and air supplies and effective personnel hygiene and practices. All of these factors should be clearly identified and associated control procedures implemented through the application of HACCP-based systems to help maintain the generally good safety record associated with these potentially high-risk products.

FRESH FRUIT JUICE

Fresh fruit and the juice squeezed from fruit has been part of the human diet probably since humankind first evolved. The low pH and high acid content of many fruit juices have, for a long time, been believed to confer safety on these products in relation to foodborne pathogens. However, commercially produced fruit juices have been implicated in a large number of foodborne outbreaks of illness in recent years, involving *Salmonella* and other enteric pathogens such as *E. coli* O157 (Besser *et al.*, 1993; Cook *et al.*, 1998; Boase *et al.*, 1999). The belief that the acidity of these products confers complete protection against enteric pathogens is clearly unsound. Fruit juice implicated in outbreaks has almost always been unpasteurised and it is this type of juice that is most vulnerable in relation to the survival of any microbial contaminants present. A high proportion of fruit juice produced is pasteurised, principally to allow the shelf life to be extended, which is otherwise compromised by the presence of fermentative yeast contaminants. Some juices also contain preservatives to restrict the growth of yeasts. Fresh unpasteurised fruit juice is viewed as a premium product and has a strong health association, although the increasing number of foodborne infections arising from its consumption is severely testing this perception. Fruit juice is consumed by all sectors of society often as a breakfast drink. With the exception of high-heat processed juice, all fresh and lightly pasteurised juices are sold under refrigeration and are intended for storage under chilled conditions.

Unpasteurised apple juice is referred to as cider in the USA. This should not be confused with UK cider, which is the term given to fermented apple juice. The term juice in this section strictly refers to the unfermented juice of fruit.

Description of process

Fruit juice can be made from many different types of fruit but the main ones consumed are orange and apple juice. Fruit are collected from orchards (Figure 4.5) and, unless used fresh, may be held in storage for long periods of time under chill conditions. The agricultural practices employed in the orchard have a significant bearing on the microbiological quality of the fruit. Contamination can arise from dust blown from the soil, from rain splash on the soil particularly to fruit grown on dwarf varieties of tree. It can also be contaminated from birds and insects. The use of natural fertilisers and the free roaming of animals will add to the microbiological burden present in fruit growing areas. Most fruit is collected from the tree

Process Stage	Consideration
Fruit growing ↓	Irrigation water quality Fertiliser source/type and application Grazing animals prohibited
Harvesting ↓	Personal hygiene Avoidance of fallen fruit
Transport and storage ↓	Hygiene Pest control Temperature
Receipt by processor and storage ↓	Supplier quality assurance programme Hygiene Pest control Temperature
Washing ↓	Water quality Chlorination/sanitisation Cross-contamination
Pressing ↓	Hygiene
Filling ↓	Hygiene Pest control
Storage ↓	Temperature
Retail sale ↓	Temperature
Customer	Advice

Figure 4.5 Process flow diagram and technical considerations for a typical unpasteurised fruit juice.

but a proportion may also be picked from the ground, known as drop or fallen fruit.

Upon arrival at the factory, fruit is usually washed in water, often dosed with chlorine, to remove soiling, dust, leaf debris, insects and stones. It is then usually transported using water flumes or conveyors to extractors. Extraction methods vary depending upon the type of fruit being processed. Apples are pressed/milled which results in the entire fruit being crushed to a pulp. Oranges however, are usually cut in half or at the base and the fruit is squeezed to release the juice which is collected via extraction pipes in the base of the squeezing/extraction unit. The reason for this difference is the need to avoid excessive release of oils from the peel of citrus fruits that would otherwise impart a burning sensation, when consumed. The released fruit juice usually passes into a buffer tank from which it is filled into containers. Depending on the product, fruit pulp may be filtered out, but for fresh juice some pulp is often left to add to the perceived 'freshness' of the product. Fruit juice may be pasteurised or it may have preservatives added prior to filling. Because of the transportation cost associated with fresh fruit or fresh whole fruit juice, many pasteurised juices are made from fruit juice concentrates which are produced in the fruit growing country and shipped frozen to importing countries where they are reconstituted with water and then re-pasteurised.

Fruit juice has a low pH (*c.* 3.5–4.5), a high water activity (*c.* 0.98–0.99) and contains a large amount of fermentable sugar. Fruit juice is sold in cardboard cartons and plastic bottles and is refrigerated. It is allocated a shelf life ranging from 5–10 days for fresh juices up to 21–30 days for pasteurised juice. Fruit juice rapidly succumbs to yeast spoilage during its life and whilst it is possible to reduce/retard this problem in pasteurised juice it is not possible to eliminate yeasts from fresh, unpasteurised juice and therefore all such juice will ultimately succumb to yeast spoilage. Yeast spoilage usually results in the production of large quantities of carbon dioxide gas (CO_2) which makes the product 'fizzy' and/or expands (blows) the pack. In addition, the ethanol and organic acids produced through fermentation impart a spoiled flavour. Pasteurising juice overcomes most of the spoilage issues associated with yeast due to their destruction in the heat process, but once the pack is opened it is vulnerable to re-contamination in the consumer's home and subsequent growth and spoilage by yeasts and lactobacilli.

Raw material issues and control

The main, and in the case of freshly squeezed juice, the only raw material used in the production of fruit juice is the fruit itself. The principal varieties of fruit used for juicing are oranges and apples. Fruit are grown in orchards throughout the world but the varieties used in premium orange juice in the UK originate from Florida (USA), Israel and Spain. The growing practices and conditions in the orchard have a significant influence on the microflora of the fruit that is delivered to the processing facility. Fruit can become contaminated with pathogens from a wide variety of sources, including the air, the soil, animal wastes and airborne creatures and insects. While it is not possible to prevent birds occasionally defecating on fruit or insects from carrying pathogens, it is possible to reduce the opportunities for enteric pathogens to be present in the vicinity of the fruit and to prevent transfer via many of the other routes.

In order to ensure sufficient nutrients are provided to the fruit trees, the farmer must supplement the soil with fertilisers. This is most commonly done using artificial fertilisers but it is a widespread practice to apply animal waste to land for agricultural crops as these not only provide an extremely useful source of nutrients but also help to maintain the condition of the soil. Clearly there are strong environmental benefits in recycling waste onto land to be used for the cultivation of crops. However, animal wastes are known to harbour enteric pathogens such as *Salmonella* and pathogenic *E. coli* and these can survive for extended periods in faeces and in the soil. In order to reduce the potential for animal wastes to be a source of pathogens in orchards, it is important to employ some treatment or agricultural control that can reduce the number of contaminating pathogens in the waste prior to application to the soil.

The greatest hazard is presented by the application of untreated waste slurries. Effectively composting the animal waste prior to application to the land can significantly reduce the hazard. Composting waste can generate very high temperatures reaching $c.\ 60°C$ or more in the centre of the pile due to the metabolic activity of the microorganisms that develop. In order to achieve effective reduction in the microbiological pathogen burden it is important to regularly turn the pile so that all components reach 'pasteurisation' temperatures. Based on calculated D values (the time to achieve a 90% reduction in the organism) in moist foods, a 6 log reduction in most *Salmonella* serotypes will be achieved at $60°C$ after only $c.\ 15$–20 minutes.

Salmonella can survive for very long periods in slurries and animal faeces. Nicholson *et al.* (2000) summarised reports demonstrating survival of *Salmonella* for up to 968 days in soil and although levels will reduce after application to soil, this is highly dependent on exposure to UV from sunlight, the dryness of the soil and a variety of other factors. Such factors have been researched but are still poorly understood and generally not under significant direct control, e.g. sunlight, rainfall, etc. Indeed, some apparent controls do not necessarily benefit product safety. For example, waste that is dug into the soil may present less of a risk of cross-contamination but this could enhance the survival of the organism due to protection from the harsher conditions on the surface.

In relation to the application method for animal waste fertilisers, as the fruit tree is already present in the soil it is not possible to plough the material into the soil and it is therefore applied directly to the surface or through subsurface injection. The longer the time between the application of wastes and harvesting of the fruit, the greater the reduction of any enteric pathogens present is likely to be. Therefore, animal waste fertiliser, if used, should be applied as soon after harvesting the previous crop as possible and should never be applied once the fruit is set and exposed. Clearly, the best practice would be to reduce the opportunity for enteric pathogens to be present in the orchard by the use of artificial fertilisers or animal wastes that have been subjected to proper composting.

Nicholson *et al.* (2000) comprehensively reviewed published work relating to farm manure applications to land and assessed the risks of pathogen transfer into the food chain. They reported that *Salmonella* had been found in animal wastes at levels up to 10^6-10^8/g in cattle faeces to 10^{10}/g in infected cattle faeces and 10^4-10^7/g in poultry manure. They reported studies demonstrating the survival of *Salmonella* in stored cattle slurry for 11–41 weeks, which varied according to pathogen serotypes, slurry composition and time of year. In dilute slurry (1–2% dry matter) survival of *S.* Dublin was estimated to be 70–80 days and in thicker slurry (7–8% dry matter), *c.* 120 days.

Salmonella survived for up to 300 days in soils spread with cattle slurry and in soils treated with animal faeces *Salmonella* survived for up to 259 days (Nicholson *et al.*, 2000).

Himathongkham *et al.* (1999a) studied the survival of *S.* Typhimurium in fresh cow manure and cow manure slurry. Samples were inoculated with the organism and stored in plastic bags in incubators at 4°C, 20°C and 37°C. In the case of the cow manure, the bags were left open which allowed some drying. Decimal reduction times derived from this work are shown in Tables 4.23 and 4.24.

Aeration of animal waste slurry storage tanks is known to have a significant effect on the reduction of enteric pathogens. Heinonen-Tanski *et al.* (1998) demonstrated that aeration of farm-scale cattle slurry tanks reduced *S.* Infantis from 300/g to less than the test detection limit (0.03/g) in 10 days in October and 460/g to 5/g in 20 days in spring months. *Salmonella* Typhimurium was reduced in pig slurry from 210/g to less than the test detection limit in 26 days (November to December).

Himathongkham *et al.* (1999b) studied the survival of *S.* Typhimurium and *S.* Enteritidis in poultry manure adjusted to different water activities. At a_w levels below 0.89 a thousand-fold decrease was observed over an 8–9 hour

Table 4.23 Reduction of *S.* Typhimurium in cow manure stored in open plastic bags in an incubator at three different temperatures, adapted from Himathongkham *et al.* (1999a)

Temperature (°C)	Location*	Ammonia (%)		pH		Moisture (%)		Decimal reduction time† (days)
		Start	End	Start	End	Start	End	
4	top	0.02	0.04	7.42	8.84	87.6	79.6	12.70
	middle		0.02		7.26		85.4	20.33‡
	bottom		0.02		7.10		85.8	
20	top	0.02	0.06	7.42	8.97	87.6	59.3	24.69
	middle		0.09		7.39		86.6	9.36‡
	bottom		0.10		7.17		86.2	
37	top	0.02	0.07	7.42	9.47	87.6	11.9	8.36
	middle		0.06		8.73		87.6	1.73‡
	bottom		0.04		8.54		88.6	

* Stack consisted of an 8-cm deep layer of cow manure (120 g) in an open plastic bag that was sampled at the top (0.5 cm below surface), middle (centre of stack) and bottom (very bottom of the bag).
† Time to achieve a 90% or 1 log reduction in the starting level of the organism.
‡ Average of middle and bottom.

Table 4.24 Reduction of *S.* Typhimurium in cow slurry* stored in plastic bags in an incubator at three different temperatures, adapted from Himathongkham *et al.* (1999a)

Temperature (°C)	Ammonia (%)		pH		Decimal reduction time† (days)
	Start	End	Start	End	
4	0.02	0.05‡	7.45	7.95‡	16.42
20	0.02	0.03‡	7.45	8.58‡	12.69
37	0.02	0.03§	7.45	8.51§	2.37

*Slurry prepared by mixing fresh cow manure with deionised water at the ratio of 1:2.
† Time to achieve a 90% or 1 log reduction in the starting level.
‡ Day 60.
§ Day 19.

storage period at 20°C. At a_w values of 0.93 and above, survival and, in one case, slight growth were observed over this time period. Extended storage of a_w adjusted poultry manure contaminated with *S.* Enteritidis was used to determine the time necessary to achieve a one million-fold reduction in the organism. The fastest time to achieve this reduction was found at a_w 0.89, which required only 8 days. At a_w 0.97 and 1.0, the time to achieve this reduction increased to 20–30 days. At a_w 0.75, 0.5 and 0.38, this reduction was again achieved within *c.* 20–30 days.

Nicholson *et al.* (2000) made provisional recommendations that slurries should be stored for periods of at least one month and preferably three months prior to application to land used to grow food. Solid manures stored for at least three months together with active turning of the manure to encourage the development of temperatures of at least 55°C were recommended prior to land spreading. Due to the increased shedding of some pathogens by some stock (young) animals, consideration should be given to handling manures from young animals separately to that from older animals and particularly ensure that they are stored for long periods or composted prior to use.

Manures should never be applied directly to ready-to-eat crops and an interval of at least six months should be observed between manure spreading and harvesting of the crop. Such recommendations, if adopted, could clearly reduce the risk of pathogens being transferred to primary agricultural crops from animal wastes whilst continuing to preserve the use of this valuable waste material.

Other factors are much more difficult to control. Water can be a significant source of contamination to fruit and vegetables in the field. Rainwater almost always has to be supplemented by other sources of water which may include local streams, rivers, wells or the mains water supply. It is incumbent on the farmer to ensure that the water supply used for irrigation is not contaminated with faecal matter. This may be through microbiological testing or from local knowledge of water sources and waste outflows. It is recognised that this is not a simple matter as water can become contaminated from a wide range of sources including agricultural run-off at points beyond the control of the individual grower. Nevertheless, as water is artificially supplied regularly to fruit trees, and contaminated irrigation water can be a major source of enteric pathogens, it is essential that the fruit growers do all in their power to avoid the use of known contaminated water sources.

One of the main factors demonstrated as contributory to outbreaks of enteric pathogen infection associated with unpasteurised fruit juice is the practice of using drop fruit in its production. As this fruit will not be consumed directly and the visual quality is not evident to the consumer, it is inevitable that drop fruit may be collected and used to make fruit juice. It is natural not to want to waste food, but it is a false economy to use material of this nature if such fruit can be the source of enteric pathogen contamination. Drop fruit can be readily contaminated from the soil. If animal wastes have been applied to the soil then the chance of the drop fruit being contaminated by enteric organisms is higher than fruit collected directly from the tree. It is also known that many farmers allow free roaming animals, including sheep and cattle, access to orchards to keep the grass and weeds down naturally. The deposition of faeces by such animals is indiscriminate and they show no regard to the potential microbiological safety implications of defecating under a fruit tree! The combination of the application of untreated waste to orchard soil, access to orchards of free roaming animals (domestic or wild) and the collection and use of drop apples is a potentially dangerous one in relation to fruit juice production. Although the incidence of contamination of fruit with bacterial pathogens does appear to be low it is apparent that it occurs with sufficient frequency to result in occasional outbreaks of serious illness from the consumption of fruit juice made from these contaminated raw materials.

Purchasers of fresh fruit for fruit juice production should stipulate the exclusion of drop fruit in the supplier specification. Although it is not cost effective to employ routine microbiological monitoring of the incoming fruit, it is certainly possible to employ visual inspection to detect excessively soiled fruit against specified limits.

In a survey of fruit producers in one state of the USA, fruit growing and harvesting practices were reviewed (Wright *et al.*, 2000a) (Table 4.25).

Table 4.25   Summary of 42 apple juice manufacturers' practices in Virginia, USA, adapted from Wright *et al.* (2000a)

Practice	Percentage of producers employing the practice
Grow own apples	67
Use manure in orchard	8
Allow animals to graze in orchard	5
Animals present in adjacent fields	54
Use drop apples	32
Use damaged apples	37.5
Wash apples prior to juice extraction	93
Use detergent-based fruit wash	18
Use sanitiser after washing	37
Brush apples whilst washing	64
Pasteurise juice	22
Add preservatives	12
Clean and sanitise equipment and facilities daily	100
Conduct microbiological testing on finished product	2
Operate HACCP	25

Of the 42 processors, 67% grew their own apples. Only 8% reported using manure to fertilise the orchard soil, 5% allowed access to free roaming animals, although animals were present in 54% of adjacent fields. Perhaps of more concern were the 32% who reported collecting drop fruit, although this was far less than the 100% who used drop apples in a survey of New England fruit growers several years before (Besser *et al.*, 1993). It is clear that practices in some orchards are improving, but more needs to be done to achieve improvements across the whole industry to limit the opportunity for contamination of fruit with enteric pathogens. Until such a time, fresh, unpasteurised juice extracted from fruit obtained from orchards in which appropriate controls are not applied will remain vulnerable to contamination and, therefore, a potential hazard to health.

Another factor likely to be of concern to this industry is the 'internalisation' of pathogens, i.e. pathogens found inside the fruit. It is self-evident that contaminants can enter the fruit through physical cuts and abrasions and there are reports of organisms residing in the stomata of plants. However, research has shown that pathogens may also be able to gain access to the inside of the fruit through the scar tissue where the stalk separates from the fruit and also at the base (blossom end) of the fruit. As

early as 1995, Zhuang *et al.* (1995) demonstrated that high numbers of
S. Montevideo could enter tomato fruit when washed in water that was
15°C colder than the fruit.

Buchanan *et al.* (1999) studied the effect of washing apples in warm and
cold water on the internalisation of *E. coli* O157:H7. Four varieties of
apple were equilibrated to 2°C and 22°C and then immersed for 20 minutes in a 1% peptone solution at 2°C containing 3×10^7 cfu/ml *E. coli*
O157:H7. Apples were then air dried in a safety cabinet at room temperature for 30 minutes and then either sampled directly or sampled after
a further washing stage. This washing stage involved immersing the apples
in a sodium hypochlorite solution (2000 ppm) for one minute followed by
draining, immersing in tap water at 22°C for one minute and then air
drying. For microbiological sampling, each apple was sectioned into eight
pieces; the skin, outer core (a section surrounding each end of the core,
the top and the bottom, to a depth of 1.5 cm), inner core and four pulp
sections (the flesh after the skin, outer cores and inner core have been
removed). Samples were homogenised and then the numbers of target
organism enumerated (Table 4.26).

In a further experiment, the effect of temperature on the uptake of a
coloured dye into the inner core region of the apple was assessed. Cold
apples (4°C) were immersed in a dye solution (20°C) and warm apples
(22°C) were immersed in cold dye solution (9°C). No dye uptake into the
inner core region was noted when the cold apple was immersed in the
warm dye solution. However, when warm apples were immersed in the
cold dye solution, 6.2% were found to have taken in a substantial amount
of dye into the inner core. A further 9.7% were found to have taken up a
slight to moderate amount of dye (Table 4.27). It was reported that dye
uptake into the inner core was always through the blossom end of the
apple and this appeared to occur where there were open channels from
the surface to the core. Dye uptake was also observed through the apple
skin in areas where there were obvious signs of bruising or damage (skin
punctures). A similar finding was reported by Merker *et al.* (1999) who
found that dye uptake into grapefruit and oranges was generally greater
when warm fruit was placed into cold water, although they also reported
dye uptake by the grapefruit on occasions where there was no temperature difference. The researchers found that dye was generally taken up
through the stem scar on the fruit but it was also found to be taken up
through old puncture wounds that had apparently healed. In general,
grapefruit seemed to be more susceptible to dye uptake than oranges.

The extent to which this may actually occur in nature is unclear but it has

Table 4.26 Uptake of *E. coli* O157:H7 in intact Red Delicious apples, adapted from Buchanan *et al.* (1999)

Apple temperature (°C)	Peptone water immersion temperature (°C)	Treatment post-immersion*	Count of *E. coli* O157:H7 (range, log cfu/g)			
			Skin	Pulp	Outer core	Inner core
22	2	No rinse	4.23–4.33	3.14–3.26	5.47–5.89	3.34–4.31
		Chlorinated water rinse	<1.00–1.30	<1.00–1.30	3.69–4.17	<2.00–3.23
2	2	No rinse	4.12–4.76	3.24–3.66	5.74–6.04	4.00–4.24
		Chlorinated water rinse	1.95–2.43	<1.00–1.60	4.12–4.57	<2.00–2.78

* After immersion for 20 minutes in peptone water containing 3×10^7 cfu/ml *E. coli* O157:H7, apples were air dried for 30 minutes and then either sampled with no further rinse or immersed in chlorinated water (2000 ppm) for 1 minute, drained and then immersed in tap water for 1 minute and dried.

Table 4.27   Effect of temperature differential on uptake of coloured dye into the inner core of apples, adapted from Buchanan *et al.* (1999)

Temperature of apples (°C)	Temperature of dye solution (°C)	Apples taking up the dye into the inner core* (%)		
		No uptake	Slight to moderate uptake	Substantial uptake
22	9	84.1	9.7	6.2
4	20	100	0	0

*A total of 113 apples were tested in each of the two treatment combinations.

potential significant implications for all types of fruit as subsequent washing and treatments of the external surfaces of fruit may be of only limited benefit if contaminants can be protected from such processes by internalisation. This merely provides further support for the need for effective controls to be implemented in the growing and harvesting of the fruit in the first place so that contamination with enteric pathogens is avoided during these stages.

Other raw materials used in the production of some fruit juices such as preservatives like sodium benzoate are not likely to contribute enteric contaminants but any juice made up from concentrate requires water. Any water used to dilute juice concentrate must be of potable quality and this should be routinely checked for bacterial indicators of contamination such as coliforms and *E. coli.*

Process issues and control

Processing of fruit to produce fruit juice is relatively straightforward and, in the absence of a pasteurisation stage, it is difficult to achieve a large reduction in contaminating pathogens if present in high numbers on the incoming fruit. Notwithstanding the fact that pathogens may on occasion become internalised in fruit, some of the process stages, in combination, can reduce low levels of incoming pathogens sufficiently to, under normal circumstances, render these products safe to consume.

Fruit may be used freshly picked or be stored for long periods (weeks to months) prior to use. This is done under chilled conditions and has little implication for the safety of fruit juice in relation to enteric pathogens.

Most processors of fruit juice subject the fruit to a washing stage prior to juice extraction. Washing has the potential to reduce contaminants present on the surface of the fruit but the level of reduction will vary according to the method of washing. Some processors use water without any added disinfectant and, whilst plain water washing can reduce the level of microorganisms on the surface of fruit and vegetables, the reduction will generally be in the order of only 1 log unit (Adams *et al.*, 1989). Indeed, the use of plain water for batch washing of fruit may actually serve to spread contaminants, originally localised on one or two fruit, across the entire batch as the organisms get washed into the water and transferred to the other fruit (Beuchat, 1992). It is generally advocated that wash water for fruit should be dosed with a disinfectant such as hypochlorite. Levels of free chlorine of *c.* 10 ppm will reduce levels of bacterial contaminants in the water. However, the effect on organisms actually located on the fruit will be limited and levels in excess of 100 ppm, which are commonly used achieve, at best, a reduction of 1–2 log units. Importantly however, providing the free chlorine level is maintained effectively, this can prevent contaminants transferring to other fruit in the batch via the wash water itself.

Beuchat *et al.* (1998) studied the effect of spraying chlorinated water onto apples and tomatoes that had been inoculated with a six-strain mixture of *Salmonella.* Inoculated apples and tomatoes, dried for 18–24 hours at 22°C, were either rinsed in tap water to simulate consumer washing or sprayed with tap water or chlorinated water (2000 ppm chlorine), left for 1, 3, 5 and 10 minutes and then rinsed in sterile water. Surviving *Salmonella* were enumerated by a rinse technique applied to the fruit (Table 4.28).

Rinsing in tap water alone achieved *c.* 0.5 log reduction in *Salmonella* on apples and 1.6 log reduction on tomatoes. Spraying with tap water and leaving for 1–10 minutes followed by rinsing in water resulted in a 1.46–1.67 log reduction on apples and 1.74–1.86 log reduction on tomatoes. Spraying with chlorinated water and leaving for 1–10 minutes prior to a water rinse resulted in a 2.08–2.86 log reduction on apples and 3.44–3.69 log reduction on tomatoes. There was no significant difference in reduction obtained by extending the treatment time from 1 to 10 minutes. The reduction achieved by water spraying alone was attributed more to the stress on *Salmonella* associated with the culture drying process followed by exposure to the soak water than to the spraying process (Beuchat *et al.*, 1998).

It is probable that the method of washing fruit is as important as the disinfectant used in the wash water. Positive methods of agitation, e.g. forced air water baths (Jacuzzis), will improve removal of pathogens from

Table 4.28 Effect of spray application of chlorinated water on the reduction of *Salmonella* on apples and tomatoes, adapted from Beuchat *et al.* (1998)

	Soak time (minutes)	Levels of *Salmonella** (log cfu/cm^2)			
		Initial level	Treatment		
			Water rinse†	Water spray‡	Chlorinated water spray (2000 ppm)§
Apples	1	4.47	3.98	2.88	2.39
	3			2.84	2.07
	5			3.01	2.04
	10			2.80	1.61
Tomatoes	1	4.53	2.93	2.67	0.84
	3			2.79	0.85
	5			2.70	1.09
	10			2.67	0.94

*Mixture of *S.* Agona, *S.* Enteritidis, *S.* Gaminara, *S.* Michigan, *S.* Montevideo and *S.* Typhimurium.
† Treated by rinsing the inoculated product prior to examination.
‡ Treated by spraying with water, leaving for specified time, rinsing in sterile water and then examined.
§ Treated by spraying with chlorinated water, leaving for a specified time, rinsing in sterile water and then examined.

the fruit and into the water where they can be decontaminated effectively. Different fruit can physically tolerate different forms of cleaning. Citrus fruit, with the thicker more resilient skin, can be subjected to more intense cleaning methods with perhaps more aggressive chemicals.

Pao and Davis (1999) studied the effect of hot water treatment and a number of chemical treatments to reduce *E. coli*, an indicator of enteric contamination, on oranges. *Escherichia coli* was inoculated onto the fruit by a 15-minute immersion in a culture followed by draining and drying at ambient temperature for two hours. Initial levels of contamination ranged from log 5.4 cfu/cm^2 on non stem-scar areas to log 5.6 cfu/cm^2 on stem-scar areas. Hot water treatment involved immersion of the fruit for 1–4 minutes in water heated to 30°C, 60°C, 70°C and 80°C. Fruit was also treated in a variety of chemical solutions for up to eight minutes. The chemicals included: sodium hypochlorite (200 ppm chlorine), chlorine dioxide (100 ppm), acid anionic sanitiser (containing phosphoric acid, dodecyl-benzenesulfonic acid and isopropyl alcohol) (200 ppm), peroxyacetic acid (80 ppm) and tri-sodium phosphate (2%).

Immersion of fruit in chemical solutions at *c*. 30°C for up to eight minutes resulted in a reduction of *E. coli* ranging from log 1.8 cfu/cm^2 (tri-sodium phosphate) to log 3.1 cfu/cm^2 (chlorine dioxide) at the non stem-scar region and only a log 1 cfu/cm^2 reduction at the stem-scar region. Interestingly, plain water (30°C) reduced the *E. coli* count by log 2.0 cfu/cm^2 and log 0.6 cfu/cm^2 at the non stem-scar and stem-scar areas, respectively, which was similar to that achieved using chlorinated water (200 ppm chlorine), although the levels of *E. coli* remaining viable in the actual wash water was not reported.

In contrast, hot water immersion achieved a 5 log cfu/cm^2 reduction in *E. coli* at both the stem-scar and non stem-scar regions after treatments of 80°C for 2 minutes and 70°C for 4 minutes. Application of these treatments, however, did result in an increase in temperature of the orange that was reported to result in a significant difference in flavour of the extracted juice between the treated and untreated fruit.

Wright *et al.* (2000b) studied a further range of sanitisers for washing apples inoculated with *E. coli* O157 by immersion and then dried in a laminar airflow cabinet for 30 minutes. All treatments were applied for up to two minutes followed by a 10 ml distilled water rinse and draining for 30 minutes prior to examination. They found that initial levels of contamination (log 3.43 cfu/cm^2) were reduced to log 2.79 cfu/cm^2 after washing in water and to log 1.37 cfu/cm^2 after treatment with 200 ppm sodium hypochlorite (Table 4.29). The greatest reduction was achieved after treatment with acetic acid (5%) which resulted in a reduction from initial levels to log 0.10 cfu/cm^2. No attempt was made to differentiate the

Table 4.29 Reduction of *E. coli* O157:H7 contamination on apples subjected to a variety of immersion treatment processes for two minutes,* adapted from Wright *et al.* (2000b)

Treatment process	Levels of *E. coli* O157:H7 (log cfu/cm^2)
None	3.43
Distilled water	2.79
200 ppm sodium hypochlorite (pH 5)	1.37
Acetic acid (5%) followed by hydrogen peroxide (3%)	1.07
Phosphoric acid fruit wash (0.3% phosphoric acid)	1.11
Acetic acid (5%)	0.10
Peroxy-acetic acid (80 ppm)	0.67

* All treatment processes lasted 2 minutes except the combination of acetic acid (5%) followed by hydrogen peroxide (3%), which were applied for 1 minute each.

reductions achieved at the stem or blossom end of the apples from those on the intact skin.

It is important to recognise that chlorine is readily inactivated by organic material and therefore regular monitoring and replenishment of the chlorine must be conducted to ensure maximum efficacy is maintained against microorganisms.

Most fruit washers have a brush system incorporated to help clean the surface of the fruit by physical action. The brushes can become a source of contamination as they are difficult to clean effectively. It is essential that brushes are regularly cleaned and sanitised and replaced when they are worn or evidently beyond repair.

Fruit is usually transported on conveyors to the automatic extraction units that vary in operation according to the fruit. Oranges enter an extraction system that cut them in half and push them onto a juicer to release the juice or, in more advanced units, the entire fruit is held in place by steel 'fingers', a circular incision is made around the base of the fruit and then the extraction 'fingers' squeeze the fruit which releases the juice into extraction pipes. This latter design avoids excess citrus oils entering the juice that would otherwise impart an unpleasant flavour to the product and cause a burning sensation when consumed. Apples, on the other hand, are pressed or milled completely to release all of the juice. It has been suggested that the two forms of processing, squeezing and pressing, have different risks associated with them in relation to the potential transfer of pathogens from contaminated fruit to the extracted fruit juice. It is argued that as orange juice is extracted by squeezing the fruit, fewer contaminants from the skin will enter the process and, whilst this view clearly has some logic, there is little evidence that this makes a huge amount of difference to the microbiological loading of the final product.

Fruit juice is collected via extraction pipes into receiving vessels. The juice may have the pulp removed, although freshly squeezed, unpasteurised juice contains a significant amount of pulp, often seen as an indication of freshness. The product is filled into cartons or plastic bottles and then stored at $< 5°C$.

Although contamination of the fruit is recognised as an important source of pathogens in fresh fruit juice, investigations of several outbreaks of illness have also implicated poor standards of factory biosecurity and cleaning as major contributory factors. It is possible that the low pH and high acidity of fruit juice has led some processors to view the product as fairly low risk in terms of microbiological safety. Consequently, the operational practices in the factory may not be consistent with the expected standards required for the manufacture of a ready-to-consume product. Such views could not be more misguided. Commercially produced, unpasteurised fruit juice products must be regarded as high-risk foods and the production facilities constructed accordingly. Factories must be built to ensure that pests cannot enter the building. During the investigation of an outbreak in Florida, USA, the presence of frogs in the factory environment was reported (Cook *et al.*, 1998) and this should clearly not be allowed to occur. Amphibians and reptiles are known to be frequently contaminated with *Salmonella* and will present a significant risk to the production environment. The building fabric must be secure with no gaps capable of allowing the entry of such pests. Windows must be screened and walls sealed to floors. Pest control agencies should be employed to provide expert advice on the control of such vectors of microbial contamination.

In addition to biosecurity, effective cleaning and sanitisation of the equipment and processing plant is critical to the safety of the product. Proper attention to these areas is equally important to product quality and shelf life. There is no doubt that the better the hygiene and cleaning of product contact surfaces is maintained in a fruit juice plant the lower the loading of yeasts in the final product and the longer the keeping quality of that product. This was ably demonstrated during the investigation of an outbreak of salmonellosis in Florida where levels of general microflora and indicators of enteric contamination were found to be much lower after more attention was given to cleaning and sanitisation of conveyors, juice extraction and filling equipment (Parish, 1998).

Areas that are known to be difficult to clean effectively, such as conveyors, extraction pipes, presses, bulk tanks and filling heads, are all items that should be targeted specifically for detailed and thorough cleaning and supported by routine visual inspection and microbiological monitoring of cleaning efficacy. The use of tests employing ATP-bioluminescence methods in kits designed for hygiene monitoring can be of real benefit in these situations as fruit juice contains very high levels of ATP and its presence after cleaning gives an indication of cleaning efficacy.

Some fruit juice is pasteurised to reduce the level of viable contaminants, yeasts in particular, allowing longer shelf lives to be allocated to the product. Temperatures usually achieve *c.* 70°C–80°C for a few seconds, which, at the low pH of fruit juice, will readily destroy even high levels of enteric pathogens, if present. However, the benefits of the heat process can be readily lost if procedures are not in place to ensure that microbial contaminants are prevented from being re-introduced during the post-pasteurisation filling operation.

Some products have preservatives added to inhibit the growth of yeasts, if present, and allow the extension of the product's shelf life. The principle preservative used is sodium benzoate, however, it is not used in many fresh, unpasteurised juices as it is contrary to the fresh perception of this premium product. Preservatives such as benzoate have some growth inhibitory effect on enteric pathogens but little or no destructive effect during chill storage.

A number of researchers have conducted investigations to determine combinations of processes capable of achieving greater reduction of *Salmonella* in finished products using freeze/thaw cycles or periods of storage at ambient temperature.

Uljas and Ingham (1999) studied the survival of a three-strain cocktail of *S.* Typhimurium inoculated at log 6.0–6.7 cfu/ml into apple juice at pH 3.3, 3.7 and 4.1. The apple juice was subject to the following processing conditions in a number of combinations; storage at 4°C, 25°C or 35°C for periods up to 12 hours, and/or freeze/thawing (-20°C for 48 hours then 4 hours at 4°C) with or without added organic acids (0.1% lactic acid or 0.1% sorbic acid). Treatments capable of achieving a reduction in the numbers of *S.* Typhimurium of > or = 5 log units in apple juice at pH 3.3 included storage at 25°C for 12 hours or at 35°C for 2 hours or after applying a freeze/thaw cycle alone. In apple juice at pH 3.7, a > or = 5 log reduction was achieved after the freeze/thaw cycle or by storage at 35°C for 6 hours. At pH 4.1, storage at 25°C for 6 hours or 35°C for 4 hours both followed by freezing and thawing was required to achieve the 5 log reduction.

A slight increase in lethality was reported when 0.1% sorbic acid was added to the apple juice. Six hours storage at 4°C or 2 hours storage at 35°C both followed by freezing and thawing achieved a > or = 5 log reduction when the juice contained sorbic acid as did 6 hours storage at 35°C. Lactic acid was reported to have no effect or in some cases actually enhanced survival of *S.* Typhimurium, e.g. during 12 hours storage at 4°C.

Organic acids are known to destroy vegetative contaminants more effectively at warm temperatures, i.e. > 25°C, although the mechanism for this effect is not clearly understood.

Final product issues and control

Fresh, unpasteurised fruit juice is sold under chilled conditions with a shelf life of 5-10 days. Pasteurised varieties are allocated shelf lives of 14 days to > 30 days depending on the level of asepsis employed in the juice filling process. Most juices succumb at some point to yeast spoilage. Fruit juice will keep longer if kept at deep chill temperatures, e.g. 1-2°C but, without freezing, temperatures below 4-5°C are not practically achievable in most consumer refrigerators.

Contrary to the belief that low pH will cause the destruction of enteric pathogens, it is clear from the outbreaks that occur, that sufficient cells can remain viable to cause foodborne illness. This is most likely to be facilitated by storage at chill temperature, which is known to protect vegetative microorganisms from the effect of organic acids. Parish *et al.* (1997) studied the survival of four *Salmonella* serotypes in orange juice adjusted to pH 3.5, 3.8, 4.1 and 4.4 at storage temperatures of 0°C and 4°C. Orange juice samples were inoculated with 10^6 cfu/ml of each serotype and their survival was quantified over defined time intervals until the level decreased to < 0.2 MPN/ml. At 0°C, the time required to achieve this reduction of *Salmonella* ranged from 14.3+/−0.9 days for *S.* Typhimurium in orange juice at pH 3.5 to 72.7+/−6.8 days for *S.* Hartford in pH 4.4 orange juice (Table 4.30). At 4°C, survival time ranged from 15.0+/−0

Table 4.30 Survival of *Salmonella* serotypes in orange juice at different pH values and stored at 0°C or 4°C, adapted from Parish *et al.* (1997)

Orange juice pH	Time (days) to achieve a reduction from 10^6 cfu/ml to < 0.2 MPN/ml			
	Temperature of storage 0°C			
	S. Gaminara	*S.* Hartford	*S.* Rubislaw	*S.* Typhimurium
3.5	15.7+/−2.5	26.7+/−4.0	18.7+/−4.5	14.3+/−0.9
3.8	24.7+/−4.5	45.7+/−6.6	39.7+/−5.9	29.7+/−2.5
4.1	45.7+/−2.6	59.7+/−3.3	48.7+/−4.8	49.7+/−3.1
4.4	60.7+/−7.9	72.7+/−6.8	63.7+/−8.7	58.3+/−3.1
	Temperature of storage 4°C			
	S. Gaminara	*S.* Hartford	*S.* Rubislaw	*S.* Typhimurium
3.5	15.7+/−2.5	24.3+/−3.4	22.7+/−2.6	15.0+/−0.0
3.8	22.0+/−2.4	32.3+/−3.8	39.3+/−6.1	26.7+/−0.5
4.1	45.7+/−2.6	56.7+/−3.9	44.3+/−6.8	43.0+/−7.3
4.4	67.7+/−3.4	70.0+/−8.6	60.7+/−2.9	48.3+/−4.2

days for *S.* Typhimurium at pH 3.5 to 70.0+/−8.6 days at pH 4.4 for
S. Hartford. The greatest difference in survival times was observed when
the pH of the orange juice was decreased, i.e. quicker reduction at pH 3.5
than at pH 4.4, and although there was a reported trend towards longer
survival times at the lower temperature (0°C), the difference between
survival times at 0°C and 4°C was not statistically significant. In addition to
the survival times, Parish *et al.* (1997) also reported the lag time before
Salmonella populations began to decline (Table 4.31). Excluding the
value of −0.25 days for *S.* Gaminara at pH 3.5 (0°C), the other values
varied from 0.16 days for *S.* Hartford at pH 3.5 and 0°C to 27.67 days for
S. Typhimurium at pH 4.4 and 0°C. Clearly, this latter lag time and some
others reported (Table 4.31) exceed the shelf life of all unpasteurised fruit
juices and such information additionally demonstrates the potential
hazard presented by the presence of this organism.

Due to the large number of outbreaks of enteric infection that have
occurred in the USA implicating unpasteurised orange and apple juice, the
United States Food and Drug Administration required processors of
unpasteurised juice to apply labelling informing the consumer of the
microbiological hazards associated with these products. The message
reads: 'WARNING: This product has not been pasteurised and, therefore,

Table 4.31 Lag time before *Salmonella* serotypes begin to decline in orange
juice at various pH values and stored at 0°C or 4°C, adapted from Parish *et al.*
(1997)

Orange juice pH	Mean lag times prior to decline (days)*			
	Temperature of storage 0°C			
	S. Gaminara	*S.* Hartford	*S.* Rubislaw	*S.* Typhimurium
3.5	−0.25	0.16	0.85	0.89
3.8	1.19	7.58	4.03	7.49
4.1	4.55	15.65	11.73	20.37
4.4	12.07	25.45	12.29	27.67
	Temperature of storage 4°C			
	S. Gaminara	*S.* Hartford	*S.* Rubislaw	*S.* Typhimurium
3.5	0.21	0.26	1.77	1.49
3.8	2.44	8.30	4.54	7.97
4.1	4.23	15.51	7.99	16.15
4.4	11.02	23.38	12.52	16.75

*Mean of three trials.

may contain harmful bacteria that can cause serious illness in children, the elderly, and persons with weakened immune systems' (Food and Drug Administration, 1998). This warning must be given on the labels of all packs unless it can be demonstrated that processes achieving at least a 5 log reduction in pathogens of concern, most notably *Salmonella* and *E. coli* O157, have been applied. It was accepted that these may include washing processes together with the application of full HACCP-based systems and associated controls. However, due to the reports indicating possible internalisation of pathogens, reliance on washing of fruit as a suitable total control is now questionable and has been reviewed in the USA (Food and Drug Administration, 1999e).

A final rule issued by the US Food and Drug Administration regarding the safe manufacture and processing of fruit juice was issued in January 2001 (Food and Drug Administration, 2001b). This required processors of fruit juice to adopt a HACCP approach to ensure the safety of fresh fruit juices. It re-emphasised the requirement for processors to achieve a 5 log reduction in the numbers of the most resistant microbial pathogen in the finished product, in comparison to the levels that may be present in the untreated juice. Processors of shelf-stable juices or concentrates using a single thermal destruction stage were exempt from these requirements.

The warning statement regarding the potential hazard associated with fresh pressed juice is still required for all other processors until HACCP programmes are implemented.

Although evidence indicated that pathogen internalisation in fruit could occur under certain circumstances the FDA, based upon a recommendation of the National Advisory Committee on Microbiological Safety of Food (NACMSF), considered the likelihood of this presenting a problem in citrus fruit processing to be low. Therefore, the 5 log reduction process could include the fruit washing stage for citrus fruit juice processing only. This approach however, also requires the adoption of routine microbiological testing for indicators of pathogen survival such as general *E. coli*, supplemented with tests for specific pathogens in the event of *E. coli* being detected (Food and Drug Administration, 2001c).

The safety of fresh fruit juice products is highly dependent on proper attention being given to a combination of controls and procedures applied in the processing environment including building and equipment construction, maintenance and hygiene, but, it is most affected by agricultural methods used in the growing and harvesting of the fruit. Whilst microbial contaminants cannot be excluded totally from the raw fruit they can be significantly reduced by the adoption of effective practices including the following:

- use of artificial or composted natural fertilisers
- exclusion of free roaming animals from the orchard
- use of potable irrigation water
- avoidance of the use of drop fruit

In this way it is hoped that future outbreaks associated with these premium products can be avoided thereby allowing the preservation of a product sector highly cherished by many consumers.

INFANT DRIED FOODS

It is difficult to think of a food group that holds more potential risks associated with its consumption than infant foods. Infants, who are taken in this book to include children from birth to the age of three years, have a much higher susceptibility than older children and adults to disease from a wide range of infectious agents including foodborne pathogens. This is due both to a less well developed microflora in the infant gut in the first few months after birth and an immature immune system that requires general exposure to antigens to become fully primed and operative. Infant foods have been implicated in a number of outbreaks of salmonellosis and these have included both milk powders (Rowe *et al.*, 1987; Louie *et al.*, 1993; Anon, 1997a) and dried cereal products (Rushdy *et al.*, 1998).

Infant foods constitute a major sector of the food industry that for many years has principally consisted of canned, bottled and dried foods. Infant milks, including dried milk formulations and ultra high temperature treated (UHT) liquid milk, together with a range of dried, canned or bottled cereal, savoury and dessert type products, make up a large proportion of the products in this sector. Dried foods are intended for reconstitution with water or milk prior to consumption and most are generally warmed before eating. They are not intended to be cooked as this may be dangerous to the infant through the inadvertent feeding of food that is too hot. In recent years, a significant investment has been made by industry in the development of new food product types for infants. Consequently, a more diverse range of products has and continues to be developed for this consumer group. The previous boundaries of canned, bottled and dried foods for infants are now being extended to include higher risk product ranges including dairy desserts, e.g. yoghurts, fromage frais, and chilled ready-meals. The following section focuses on the manufacture of dried infant foods.

Description of process

Dried baby foods are manufactured from an extremely wide range of raw ingredients that vary depending on the finished product type required. The largest group of foods in this sector are probably the infant dried milks that are formulated to meet the complex nutritional needs of infants in ages ranging from birth to 12 months and, in the form of follow-on milk, for those over 12 months of age. Dried savoury and dessert foods are formulated for age ranges from three months and beyond and are used as weaning foods that allow the infant to progress from liquid, homogeneous products to more particulate or solid foods. At increasing ages, the number and size of the particulates in the infant foods increases to encourage the development of solid food consumption.

Infant dried milks are made (Figure 4.6) from cows' milk that may be delivered to the drying site in an unpasteurised or pasteurised state. It is held in a buffer tank and then subjected to pasteurisation. The solids content of the milk is then increased by passing the milk through an evaporative condenser to produce milk concentrate of about 50% milk solids content. Evaporative condensers usually operate under vacuum; this depresses the boiling point of the milk, allowing water to evaporate at lower temperatures, thereby minimising the adverse effect of high heat on the nutritional and organoleptic quality of the milk. The concentrated milk is stored in buffer tanks from where the milk is fed to a drying chamber.

Milk powder is most often produced using large spray drying chambers. Milk is pumped into the top of the drying chamber through an atomiser that may be of a rotary or jet design. Milk droplets are formed when the atomised milk meets the hot air that is fed into the chamber at the same time. The size of liquid stream and the temperature, flow rate and direction of the hot air are critical factors that affect the efficiency of the spray drying process. The air, which is heated to temperatures of 180–250°C by direct gas heaters or steam heaters, dries the milk droplets, which rapidly form into milk powder particles. The milk powder settles and is cooled and conveyed out of the drying chamber to a bag-filling unit. A cyclone unit separates small and light particles from the cooling air and these are returned to the main powder stream for filling. In addition, there are usually cloth filters that trap the finest milk dust particles known as fines. Milk powder is often re-wetted with steam and re-dried on a fluidised bed; this improves the solubility of the product if it intended that the product be re-hydrated.

Dried milk powder is fed to hoppers from where it is filled into lined, brown paper sacks, usually in 25 kg quantities. These are delivered to the infant food production unit. Here the raw material dried milk powder is dry blended with other ingredients, including vitamins and minerals prior to filling into cans. A plastic measuring spoon may be placed in the can and then they are usually sealed with metal foil prior to covering with a 'snap-on' plastic lid. The products are stored and distributed under ambient conditions. Detailed customer use instructions are given with the product and these provide guidance on methods of re-hydration and warming.

Dried savoury and dessert products are manufactured from a much larger

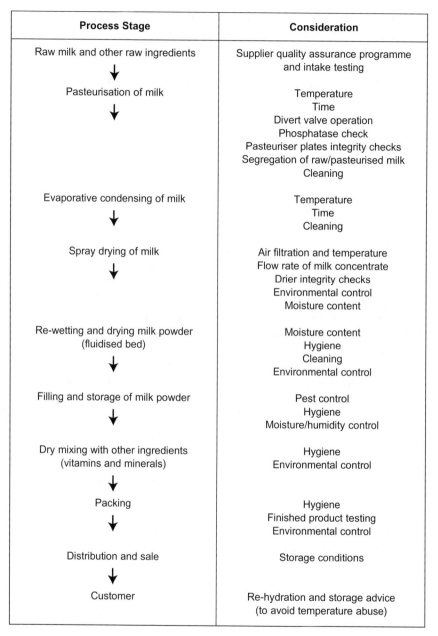

Process Stage	Consideration
Raw milk and other raw ingredients ↓	Supplier quality assurance programme and intake testing
Pasteurisation of milk ↓	Temperature Time Divert valve operation Phosphatase check Pasteuriser plates integrity checks Segregation of raw/pasteurised milk Cleaning
Evaporative condensing of milk ↓	Temperature Time Cleaning
Spray drying of milk ↓	Air filtration and temperature Flow rate of milk concentrate Drier integrity checks Environmental control Moisture content
Re-wetting and drying milk powder (fluidised bed) ↓	Moisture content Hygiene Cleaning Environmental control
Filling and storage of milk powder ↓	Pest control Hygiene Moisture/humidity control
Dry mixing with other ingredients (vitamins and minerals) ↓	Hygiene Environmental control
Packing ↓	Hygiene Finished product testing Environmental control
Distribution and sale ↓	Storage conditions
Customer	Re-hydration and storage advice (to avoid temperature abuse)

Figure 4.6 Process flow diagram and technical considerations for infant dried milk powder.

variety of ingredients. Raw ingredients such as meat and vegetables are usually pre-processed by peeling, slicing/dicing and mixing in cooking vessels (Figure 4.7). They are fully cooked to temperatures usually approaching boiling prior to being fed to roller driers. In most cases, the drier is fed with a liquid or concentrate, held in hoppers prior to being applied via a trough and several small roller drums as a thin film onto the main roller drum of the drier. Roller dried product reaches temperatures

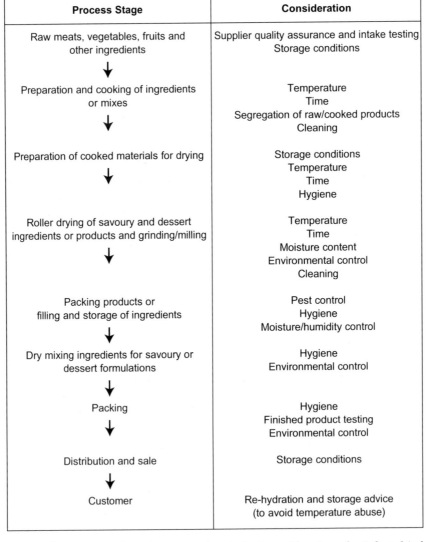

Process Stage	Consideration
Raw meats, vegetables, fruits and other ingredients	Supplier quality assurance and intake testing Storage conditions
Preparation and cooking of ingredients or mixes	Temperature Time Segregation of raw/cooked products Cleaning
Preparation of cooked materials for drying	Storage conditions Temperature Time Hygiene
Roller drying of savoury and dessert ingredients or products and grinding/milling	Temperature Time Moisture content Environmental control Cleaning
Packing products or filling and storage of ingredients	Pest control Hygiene Moisture/humidity control
Dry mixing ingredients for savoury or dessert formulations	Hygiene Environmental control
Packing	Hygiene Finished product testing Environmental control
Distribution and sale	Storage conditions
Customer	Re-hydration and storage advice (to avoid temperature abuse)

Figure 4.7 Process flow diagram and technical considerations for infant dried savoury and dessert products.

usually not exceeding 90°C. The product is scraped from the roller drier and, after some breakdown/sieving of the particles, is filled into sacks. Usually, finished products are made by thorough blending of the required dry ingredients and filling the mixture into packs. As with infant dried milks, full customer guidance is given on-pack regarding reconstitution and heating prior to consumption.

These dried products have a very low moisture content (< 10%) and are ambient stable. They are packed under conditions that exclude moisture and, in an unopened state, remain stable for 12 months or more.

Raw material issues and control

The raw materials used for milk-based, savoury and dessert dried infant foods span almost all commodity groups. In many cases the ingredients are sourced and received in the raw state including raw milk, raw meat and raw vegetables. It is not practical to discuss the entire range of ingredients used in the production of this group of products, but it is clear that food ingredients for infant consumption must be carefully sourced and selected and the suppliers should be subject to effective raw material quality assurance programmes. *Salmonella* will be present on occasion in raw meat and milk and to a lesser extent on raw vegetables. It is important, as part of a HACCP-based system, to carry out a formal assessment of the microbiological hazards presented by different ingredients to identify those ingredients that may require specific controls and monitoring procedures.

Due to the hazards associated with handling unprocessed raw ingredients at the same site the infant foods are made, some manufacturers source and utilise previously pasteurised or cooked ingredients; wherever possible, this is the preferred and safer approach. Ingredients such as herbs and spices should be sourced from suppliers using heat or other treatments to reduce the microbial load. The raw ingredient suppliers should be routinely audited to ensure adherence to effective good manufacturing practices and their products should be subjected to routine microbiological monitoring at the point of intake to the infant food production site.

In some cases, the infant food producer may buy high-risk dried ingredients directly from other suppliers, and will merely dry blend the ingredients to make the final product. It is absolutely critical that relevant controls are in place to assure the safety of the manufacture and supply of any such materials through vendor quality assurance programmes. Indeed, it is apparent that an outbreak of salmonellosis affecting eight

infants in England in 1995 caused by *S.* Senftenberg was due to a contaminated cereal used as a raw material by a baby food manufacturer
(Rushdy *et al.*, 1998). Although the organism was never isolated from a
finished product, the baby food manufacturer reported rejecting a con-
signment of the cereal from the raw material producer in 1994 as it had
arrived in a poor condition. *Salmonella* Senftenberg had been isolated
from a sample of cereal taken from this batch and although none of this
batch was used or distributed, it was apparent that the cereal supplier
used the same milling machinery for the baby food ingredients as for
other, possibly non heat-treated ingredients destined for other uses.

Whilst these foods are not chilled perishable products, the principles of
high/low-risk separation at such production sites must be in operation
between raw materials and finished products in the manufacture of infant
foods or ingredients destined for such foods. This must include separate
storage areas for ingredients that are raw and may contain pathogens,
such as meats and vegetables, and those which may be ready to eat and
intended directly for dry blending in the production of the finished
products, e.g. dry powders. Strict procedures must also be in place to
prevent personnel becoming vectors of contaminants from external
sources or through handling raw products or indeed, from infected individuals or those carrying infectious disease agents such as enteric
pathogens.

In addition to dry ingredients bought-in directly, many of the food materials used to make these products are pre-processed in the factory from
raw ingredients. Meat and vegetable mixes are, for example, cooked and
blended to form the mixture that subsequently will be dried to form the
powdered product. Storage conditions of ingredients must be appropriate
to the material concerned. Dry ingredients must be kept stored under
cool, dry, pest-free conditions and chilled foods should be stored under
effective refrigeration ($<8°C$) and ideally at $<5°C$ to preclude the potential increase in levels of enteric pathogens. Many foods are received and
stored frozen and tempering of frozen materials, which should be conducted under chilled conditions, must be carefully controlled to prevent
growth of microbial contaminants.

Equally important as the control of ingredient storage conditions is the
control of the various holding stages that are inherent in the pre-proces-
sing of these types of products. Cooked mixtures that are to be dried are
often held in buffer tanks for feeding various hoppers and driers. It is
essential to ensure that these holding stages are carefully controlled to
prevent the growth of contaminants through ineffective temperature

control or their build up as a result of ineffective cleaning of equipment. All processing stages prior to drying and their associated operational temperatures, times and cleaning procedures should be included in the hazard analysis done by the manufacturers of these products.

Process issues and control

The process employed for the manufacture of dried milk products starts with the pasteurisation of the milk itself. This needs to be conducted in accordance with the standard requirements for the pasteurisation of milk and therefore must always meet the minimum standard of 71.7°C for 15 seconds or an equivalent process (in the UK and Europe), although in the case of milk for infant dried milk the pasteurisation process is usually much higher than this minimum requirement.

The pasteurisation stage must be under effective control. Pasteurisation temperatures and milk flow rates should be continuously monitored. Such on-line monitoring systems must be linked to a flow diversion system that can re-direct milk from forward flow back into the heating phase or into a buffer tank if the specified process limits are not achieved. All equipment used for monitoring such critical elements of the process, e.g. temperature probes, must be routinely calibrated. Temperature should be recorded using continuous chart recorders and all controls should be supplemented by routine tests on the pasteurised milk for indicators of process efficacy, e.g. destruction of the enzyme alkaline phosphatase. This enzyme should be destroyed by effective pasteurisation and its presence in pasteurised milk, immediately after the heat processing stage, is an indication of inadequate pasteurisation. Robust action plans must be in place for responding to any process failures identified.

In order to produce dried powder, the milk is concentrated through an evaporative condenser prior to spray drying. Evaporative condensers are used because the economics associated with removal of water from milk favour evaporative removal rather than directly spray drying. Therefore, removing moisture from the product at this stage achieves a more cost-effective production process. The condenser, which often incorporates a re-pasteurisation stage as part of pre-heating, can also provide opportunities for microbial proliferation. Condensers are divided into a number of sections over which the milk falls. These operate under vacuum to reduce the boiling temperature of water and therefore improve the efficiency of moisture removal whilst reducing adverse organoleptic effects and nutritional deterioration of the product. The sections of the condenser operate at gradually reducing temperatures. The milk is first pre-heated to temperatures of 70–100°C and there is a gradual reduction in temperature in the first and successive sections, reaching 40–50°C in the final section of the condenser. Whilst pasteurisation of the milk occurs during the pre-heating process, any surviving microorganisms are exposed to potential growth temperatures later in the condenser. Clearly, this should not be of concern in relation to enteric pathogens as these are destroyed and should not be present at this stage. However, it could be a concern with regard to other organisms, e.g. spore-forming or thermoduric bacteria that may have survived the heat treatments, and conditions must be carefully controlled at this stage of the process to ensure that microbial hazards do not develop.

It should be recognised however, that although the temperature becomes more conducive to microbial growth during the latter stages of evaporation, the milk is also becoming more concentrated. The high solids contents approaching 50% will depress the water activity resulting in slower growth of organisms. Clearly, the effect on microbial growth of the combinations of temperature and increasing milk solids content developing through the process need to be considered in the hazard analysis carried out by each manufacturer. This analysis should also assess the potential for survival and growth of microorganisms in the food materials being processed and on/in the production equipment. Production runs should be operated taking account of the potential for build-up of contaminants and equipment should be subject to full cleaning and sanitisation at frequencies determined from the hazard analysis. In some cases this may require cleaning to be carried out several times in a production day.

After concentration, the milk is pumped to a buffer tank prior to transfer to the spray drier. As for all intermediate products, it is essential that in addition to controlling storage temperatures and times to preclude the growth of any surviving or contaminating microorganisms, the equipment used to transfer and hold material must be regularly and effectively cleaned and sanitised to remove debris and destroy residual contaminants. Concentrated milk should be sent on for spray drying quickly, otherwise, growth/build-up of contaminants may occur. Depending on the microbial growth conditions in the concentrate it may be necessary to cool the material or maintain it at a high temperature (> 65°C) if it is intended to be stored prior to further processing.

Inadequacies in the primary stages of processing in dried milk production have resulted in microbiological incidents involving both foodborne pathogens and spoilage organisms (International Commission on Microbiological Specifications for Foods, 1980).

Cleaning of equipment should be supported by routine monitoring of the cleaned equipment for indicators of cleaning efficacy such as coliforms, aerobic plate count (APC) or even aerobic spore-forming bacteria. The use of hygiene tests based on ATP-bioluminescence methods are also useful for the detection of milk residues not effectively cleaned from the holding or transfer equipment.

Powdered savoury and dessert products are not usually concentrated in such a way prior to drying. Ingredients are usually weighed into hoppers and then mixed or bowl chopped with water to the desired consistency. They are then fully cooked in processes that usually raise the mixture to boiling temperatures for a considerable time. The cooked product is then pumped to a buffer tank prior to drying or cooled and stored until required for drying. It is essential that cooking processes are fully and effectively validated to ensure that the process times and temperatures used will consistently achieve full pasteurisation of all components in the mixture. Cooking validation studies must take account of all worst-case conditions such as piece/particle size, temperature of ingoing raw ingredients and the minimum heat process. Production processes determined from the results of validation work must then be supported with appropriate automatic process controls linked to automatic monitoring devices, e.g. chart temperature recorders with properly calibrated probes, to ensure the critical factors are under control at each relevant stage of the process. These should then be checked by routine manual checks of temperature, particle size, etc. to ensure they remain under control. Effective reaction procedures to process failures must be in place.

As described for milk concentrate, intermediate dessert or savoury mixes will support the growth of any surviving spore-forming bacteria or post-process microbial contaminants. It is therefore essential to limit the time between preparing the cooked mixture and the drying process or, if this is likely to be excessive, e.g. >4 hours, to properly chill the mixture and hold it under refrigeration. Equipment used for storing and pumping the material must be routinely cleaned and sanitised to prevent any build up of microorganisms, and the efficacy of such procedures should be checked by suitable examination techniques, i.e. microbiological swabbing.

Concentrate to be spray dried is fed into a drying chamber through a jet or rotary atomiser, the former producing a thin cone of milk concentrate and the latter forming a stream of droplets. In both cases the intention is to introduce a fine stream of milk concentrate into the chamber and dry it by hot air. The air enters the chamber after being heated by direct gas/ electric burners or via steam heaters. The air inlet is designed to achieve optimal break-up of the milk stream into droplets and drying of the milk and therefore it may flow counter current or at various angles to the incoming milk stream. Hot air enters the chamber at temperatures between 180 and 250°C and, as it passes the milk droplets, it raises the temperature of the droplets and causes them to lose moisture and consequently dry. Dried milk particles may develop in various sizes depending on the flow rate of the milk and the airflow rate/temperature. The dried milk particles may either drop to the bottom of the chamber or they may flow out of the chamber with the hot, moisture-laden air into a cyclone. Exit process air temperatures can vary between 70 and 90°C. Most spray driers incorporate cyclones to concentrate the milk powder particles and ensure effective capture of all milk particles from the exiting air streams. Small particles may also be captured via a series of cloth filters that collect 'fines'. Powder is cooled by cool air prior to bagging. Powder collected at the bottom of the chamber or through cyclones, depending on its intended use, is often further processed on a fluidised bed which carries the product into a stream of steam. This re-wets the powder, which is then re-dried on the bed by hot air, then cooled by cooling air prior to bagging. This process improves the solubility of the product and avoids the formation of clumps when the consumer re-hydrates the product.

Although the air entering the drying chamber is extremely hot, the milk droplets do not achieve similar high temperatures because of the heat lost during the evaporation of the water. Clearly, this is rather difficult to measure but milk temperatures are well below 100°C, probably 70–80°C, and studies examining the effect of spray drying on microbial populations have shown that general microflora and *Salmonella* do survive the spray drying process. For example, Thompson *et al.* (1978) showed that *E. coli,* an indicator of enteric contamination, was reduced by different amounts depending on the exit air temperature of the drier. At 71.1°C, a 2 log reduction was achieved whilst at 93.3°C a >3 log reduction was noted (Table 4.32). Also, Licari and Potter (1970) studied the survival of *Salmonella* in spray-dried milk powder. Contaminated milk powder was produced by spray drying milk, inoculated with *S.* Typhimurium or *S.* Thompson, at an outlet air temperature of 79.4°C to achieve initial contamination levels of *c.* $10^4/100\,g$ milk solids. The survival of the organisms was studied by storing the powder at 25°C, 35°C, 45°C and 55°C for up to eight weeks (Table 4.33). Levels reduced in all cases but, with the exception of 45°C and 55°C storage for *S.* Typhimurium only, the organism remained detectable for the eight-week storage period.

Whilst these studies clearly show some reduction in levels of vegetative organisms during the drying processes applied, they also show significant

Table 4.32 Reduction in *E. coli* inoculated into skim milk concentrate (1×10^6/g) and spray dried using different inlet/outlet air temperatures, adapted from Thompson *et al.* (1978)

Air temperature (°C)		Total solids in milk concentrate (%)	Moisture in dried milk powder (%)	Levels of *E. coli* expected in milk powder if no destruction	Levels of *E. coli* after spray drying
Inlet	Outlet				
142.2	93.3	40	3.00	2.42×10^6	5.0×10^2
140.5	82.2	40	4.60	2.38×10^6	5.1×10^3
129.8	71.1	40	4.80	2.38×10^6	1.1×10^4

Table 4.33 Survival of *Salmonella* in dried milk powder contaminated prior to spray drying and stored at various temperatures for eight weeks, adapted from Licari and Potter (1970)

Salmonella serotype	Storage temperature (°C)	*Salmonella* levels (log cfu per 100 g milk powder)			
		Storage (weeks)			
		0	2	4	8
S. Typhimurium	25	1.64×10^4	9.45×10^2	3.57×10^2	9.45×10^1
	35	1.64×10^4	3.57×10^2	2.46×10^2	5.57×10^1
	45	1.64×10^4	1.74×10^1	0.513	<0.002
	55	1.64×10^4	<0.002	<0.002	<0.002
S. Thompson	25	9.18×10^4	5.63×10^3	3.55×10^3	2.45×10^2
	35	9.18×10^4	3.55×10^3	9.48×10^2	2.45×10^1
	45	9.18×10^4	1.12×10^2	3.55×10^1	2.4
	55	9.18×10^4	4.68	<0.002	<0.002

potential for survival of *Salmonella* and are clear evidence that the drying stage itself should not be relied upon to achieve a safe product.

Very little opportunity exists for growth of microorganisms once the milk is dried. Milk powder usually leaves the drier at a moisture level below 10% and often closer to 5%. Even during re-wetting, the product moisture level stays below 15% and little opportunity exists for growth. Care must, however, be taken to avoid build up of local pockets of moisture on the sides of the chamber in the drier itself. This can occur due to excessive milk flow rates, which results in milk particles sticking to each other and to the sides of the chamber. Such areas in which moisture is trapped and where the temperature remains warm could present potential problems once such material is dislodged or falls from the side.

In contrast to spray drying, roller drying achieves a higher lethality to microorganisms during the process because of the greater contact between the product and the heat source, the drum. Product, which is usually pre-heated to in excess of 80°C (and sometimes above 100°C), is applied as a thin film to the external surface of the drum drier via several intermediate drums and as the drier rotates the product is heated up. The inside of the drum is heated to temperatures of 140–150°C, this raises the product temperature to *c.* 90°C, well in excess of a full pasteurisation process, and moisture is driven off. In fact, one reason why dried milk powder is more commonly manufactured using spray rather than roller driers is because of the cooked flavour that roller drying can impart to the product due to the higher temperatures achieved by the process. Dried material is collected from roller driers by a scraper that is closely aligned to the drum to ensure removal of all material from its surface.

Product from spray or roller driers may go through some sieving or particle-breaking process prior to or after storage and is then packed into containers or sacks depending on its intended use.

Potential for microbial contamination of product within the spray drier itself is of great concern. It is clear that there are a number of areas in the spray drier where microbial contaminants could enter. Air should not be a source of contaminants as all air to spray driers is usually passed through full microbiological filtration systems that incorporate high efficiency particle removing (HEPA) filters and it is heated to very high temperatures. Where this is not the case or where there are failures in the air treatment system, air supplies will need to be regarded as potentially hazardous to both the process and the product. Milk concentration and handling processes have already been discussed and providing this is properly controlled should not be a source of enteric pathogens. The principal areas where microbial contaminants may enter are from the production equipment itself, including the drying chamber and associated cyclones, fluidised bed, conveyors, fines cloths (if used) and even the atomiser or jet spray nozzles. The sources of contamination for this equipment are the internal or external environment and people.

The jet or rotary atomisers are difficult to clean and are subject to extensive build up of product. It is usual to utilise several rotary atomisers and interchange these during production to allow them to be cleaned frequently. The frequent and effective cleaning and sanitisation of this equipment together with their associated pumps, all of which come into contact with all the material entering the chamber, is absolutely critical to the safety of the final product.

This also applies to all other parts of the production environment and equipment, from the drier through to the product bagging area. As the post-drier environment is a dry area, most cleaning is done using dry cleaning methods. Vacuum cleaners, brushes and alcohol-based disin-fectant wipes are commonly used to keep the equipment and environ-ment free from the excessive build up of product dust. The drier chamber is wet cleaned at intervals ranging from weekly to monthly. This is usually done using clean-in-place (CIP) systems. Many of the other areas of the plant are not wet cleaned at all except perhaps during periodic seasonal breaks that coincide with major engineering work and maintenance.

The drying chamber itself is clearly a major potential source of microbial contamination. As learned from the outbreak of salmonellosis due to infant formula milk in the UK in 1985 (Rowe *et al.*, 1987), the drying chamber integrity must be maintained and must be subject to routine and rigorous inspection. In the investigation of this outbreak, which affected 48 infants and resulted in the death of one, detailed inspection found a number of holes in the inner lining of the chamber and one large tear from where *Salmonella* entered the chamber and contaminated the drying milk powder (see Chapter 2). Such an occurrence reinforces the need for effective maintenance and control of the drier itself and its environs to ensure external sources of contamination cannot gain access to the chamber. Reliable equipment maintenance and environmental protection and cleaning programmes equally apply to the fluidised bed, conveying systems, cyclones, and all other product contact surfaces.

Indeed, this outbreak also clearly demonstrates that control must be exercised over the external environment of the powder processing plant which should be fully enclosed inside an environmentally secure factory building. The section enclosing the drying and powder packing plant should be sealed completely both from the external environment and from the low-risk side of the operation where raw materials are handled and processed. All air into the enclosed facility must be filtered, all products should be heat processed passing into the area and all personnel should enter through a high/low-risk changing facility. Entry procedures for individuals must include the removal of all external clothing and footwear. Footwear, overalls and hair covering should be worn, all of which must be dedicated to the high-risk side of the operation. Other potential sources of contamination such as water must be chemically treated or filtered to a level sufficient to destroy/remove vegetative microorganisms. Any other materials entering the facility should be treated as they pass into the high-risk area, e.g. by using a sanitiser spray on packaging and utensils leaving sufficient contact time to ensure effective destruction of any surface microbial contaminants.

Once produced, powder may be packed in 25 kg quantities in paper sacks and despatched to a further processing operation or it may be conveyed to a large storage container prior to being used for mixing with other ingredients, including vitamins and minerals, to produce infant milk formulas. These products are filled into cans or plastic-lined boxes, a measuring spoon added, and then these are sealed, lidded and despatched.

Care must be taken in the mixing and filling operations to prevent the introduction of *Salmonella* from personnel, equipment or the environment. This is achieved by implementing and maintaining the practices and procedures required for high-risk food production areas already described.

With effective control of the food material entering a drying plant, it is reasonable to assume that, if present, organisms such as *Salmonella* are most likely to contaminate the process and product only from the external environment and people. It is for this reason that significant effort is usually put into monitoring the environment around a drier and a drying plant for the presence of enteric pathogens such as *Salmonella* as well as tests for bacterial indicators of contamination such as coliforms or Enterobacteriaceae. Environmental sampling plans for monitoring the presence of *Salmonella* usually involve taking large numbers of swab and/or debris samples from floors, walls and drains in the areas surrounding both the drier and the product bagging machines and also relevant areas outside the factory and in roof spaces, if these are accessible. Filters and pre-filters from the air supply system are sampled when these are routinely changed and powder that is vacuumed from floors and taken from 'fines' cloths is also routinely sampled to provide information about the microbial integrity and safety of the plant. The assumption should be that if *Salmonella* is found present in the internal or immediate external areas associated with the drier and dried product handling it could well become a hazard in the product, therefore as much attention should be given to examining samples taken from the environment as those taken of product for the presence of *Salmonella*.

A microbiological survey of 18 dry dairy product-processing plants in the USA was carried out by Gabis *et al.* (1989) to determine the status of a number of environmental sites within the plants. Processing plants producing butter, cheese and dried milk products were included in the survey although the number of each type of plant is not given. *Salmonella* was isolated from one floor drain in a butter production area, one floor drain in a drier room and from one air pre-filter to a drier (Table 4.34).

Table 4.34 Incidence of *Salmonella* in 18 dry dairy product-processing plants, adapted from Gabis *et al.* (1989)

Sample point	Number positive/ number tested for *Salmonella*	Site positive
Raw milk receiving	0/38	
Wet processing	1/103	Floor drain in butter production area
Cheese plant	0/10	
Whey plant	0/10	
Drier room	1/28	Floor drain in drier room
Bagging room	0/15	
Air/ventilation system	1/30	Pre-filter to drier
Miscellaneous	0/11	

The risk from contamination by *Salmonella* is somewhat less in a roller drier operation than a milk powder plant because it is not subject to the same level of air movement into the unit as a spray drier, temperatures of material during drying are higher and access for cleaning is easier. However, similar procedures for equipment, personnel and environmental control should still be implemented.

Final product issues and control

The moisture content of infant dried milk powder or other dried dessert or cereal mixes is so low, *c.* 5–10%, that the potential for growth of any contaminating microorganisms is precluded. The dry state of these products must be maintained to ensure product stability as any local increase in moisture level within the product due to condensation or high humidity can cause deterioration in product quality and may allow microbial growth. In sealed packs, product is protected from ingress of moisture and the finished products are considered to be stable. However, just because the product environment is inhospitable to microbial growth, it does not mean that microorganisms will not survive in dried powders. It has been known for some time now that contaminating microorganisms including *Salmonella* can survive for extremely long periods in low water activity substrates. This ability has been demonstrated in milk powders (Table 4.33).

Ray *et al.* (1971a) studied the survival of *Salmonella* in naturally contaminated dried milk powder stored in polythene bags at room temperature. Although rates of detection reduced over time, positive samples were still found in batches stored for the full duration of the study (10–12 months). In batches of contaminated product stored for ten months from one production plant, the detection of *Salmonella* (in 100 g samples) reduced from 100% (10/10) at the start of the study to 50% (5/10) after 2 months storage and further to 15% (2/13) at the end of the study. Enumeration studies found that levels of 2.2 MPN/100 g in 2-month-old samples reduced to 0.9 MPN/100 g after a further 4 months' storage.

The hazard of enteric pathogens present in infant foods is clearly exacerbated by a number of factors. Firstly, the product is intended for an age group that is particularly susceptible to infection from low numbers of pathogenic microorganisms. This is due to the less well-developed intestinal microflora and immune system in infants. Secondly, most of these dried products are intended for reconstitution with water and heating prior to consumption and, due to the need to avoid overheating baby foods because of the potential to scald the infants' mouths, infant foods are merely warmed after reconstitution. Clearly, the prepared food should be consumed quickly after warming but this requirement is likely to be frequently abused, e.g. in circumstances where mothers need to prepare the food in advance for feeds during outings, etc. The warming of infant food is as desirable for the baby as it is for any contaminating pathogens such as *Salmonella*, but for different reasons. Any extended periods of time at the temperatures (30–40°C) achieved during warming of reconstituted dried baby foods will allow significant growth opportunities for microorganisms.

Instructions on infant food product packaging must clearly include the need to reconstitute, warm and feed quickly to the infant to ensure the product remains safe and wholesome. Whilst *Salmonella* and other vegetative bacterial pathogens should not be present in the product, other contaminating organisms including bacterial spore-formers may occasionally be present and such advice relates principally to the control of these potential contaminants.

Dried milks and infant savoury and dessert products are subjected to extensive microbiological sampling programmes for *Salmonella*. It is common to undertake a high level of sampling, e.g. 25 g (minimum) samples taken from every fifty, 25 kg bags produced, i.e. at least 0.002% of material produced. These days, sampling is often done using a continuous sampling device. A screw type auger is fitted in line at the bagging plant and this turns continuously whilst product is being filled, removing small quantities of product on a continuous basis throughout the filling process. This form of sampling results in the collection of several kilograms of material during each filling run. This large quantity is then further mixed and sub-sampled for examination. This method of sampling significantly increases the chances of capturing any microbial contaminants that may be present in a batch of product. In practice, if a microorganism is likely to be present in a product batch at a low frequency and if it is not homogeneously distributed in that batch, the chances of capturing it in a sample is exceptionally low (International Commission on Microbiological Specifications for Foods, 1986). Taking a large number of small samples across the batch will greatly increase the chances of detecting the organism if present, whereas taking a larger individual sample, i.e. 100 g versus 25 g, will have negligible effect on improving the chances of its recovery.

Few recent surveys of the incidence of *Salmonella* in infant dried powders have ever found the organism to be present as this is a rare occurrence and the chances of finding it are fairly remote even if it is present. Ray *et al.* (1971b) reported on surveys of the incidence of *Salmonella* in eight dried milk powder plants. They reported finding *Salmonella* in 5.6% of 963 product samples, 23.6% of 1125 air filter samples and 21.1% of 763 environmental samples.

In the first six months of 1967, the United States Department of Agriculture sampled both products and the environments of 200 dry milk powder plants in 19 States and found 1% of the 3315 samples of product and 8.2% of the 1475 environmental samples to be contaminated with *Salmonella*; 21 different *Salmonella* serotypes were isolated (Schroeder, 1968). The situation has somewhat improved in milk powder plants over the last three decades and the finding of *Salmonella* in a sample of milk powder in a modern plant today is unusual and, if found, would generally result in shut-down of the plant.

A survey of UK foodstuffs carried out in 1990–91 reported the detection of *Salmonella* in two out of 327 dried milk samples (Anon, 1993a).

The safety of dried infant foods is dependent on the effective implementation of controls of raw material sources and handling, milk treatment processes, equipment and environmental integrity and hygiene, personnel and final product handling practices. There is absolutely no room for error in a process whose product is destined for such a vulnerable group and it is incumbent on manufacturers of such products to operate to the highest standards of safety possible. Those not prepared to operate such standards for economic reasons should not be in business. The impact of an outbreak of salmonellosis on a business in this sector is usually dramatic and could involve the closure of the business.

CHICKEN AND EGGS

Chicken and eggs are two of the most commonly consumed foods in the UK and many other countries. They are also amongst the most common foods implicated in food poisoning outbreaks with many cases of illness being reported every year in both the UK and most other industrialised countries (Evans *et al.*, 1998; Cronin, 1999; Trepka *et al.*, 1999; Reporter *et al.*, 2000).

Over 500 million broiler chickens are produced in the UK each year and nearly 10 000 million eggs are consumed every year. To supply such demand, the scale of production is also very large. This has resulted in the intensification of farming and slaughter/processing facilities to ensure the demand is met. 'Recent' intensification of farming has often been quoted as a reason for increased incidence of pathogens, but such intensification is not recent. The stimulus for intensification can be traced back to the post-war years in the UK (1940s–1950s) when the country needed farm production processes to develop rapidly to provide adequate food supplies to the nation. Intensification methods do not cause increased incidence of pathogenic microorganisms in animals *per se*. Such methods, in fact, allow better control of sources of contamination. However, it is evident that, once introduced into an intensive farming system, pathogenic microorganisms are offered greater opportunity to spread under the intensive growing conditions.

Poultry and eggs have become synonymous with *Salmonella* in the last two decades, due to the high number of outbreaks of salmonellosis implicating such products and the heightened media and political attention given to the issue.

These products are consumed by all sectors of society. Most eggs (*c.* 80%) are produced by intensive farming systems and a smaller percentage (*c.* 15%) is produced by free-range hens. In addition to direct retail sale of fresh, raw, shell eggs (*c.* 60% of the market), a significant market exists in catering outlets (*c.* 20%) and food manufacturing processes (*c.* 20%) for which the product is often supplied in liquid form as whole egg, egg yolk or egg albumen.

For many years the largest sector of the market for chicken was in frozen, whole birds but since the 1980s, this has been overtaken by the sale of fresh, chilled, whole birds and fresh and frozen chicken portions. Chicken is also used in a wide variety of further processed products, e.g. ready-meals to be cooked before consumption or it is cooked in manufacturing plants for use in ready-to-eat products, e.g. chicken salad meals and sandwiches.

Description of process

Eggs are produced by laying hens (layers) and broiler hens are used to provide chicken meat, however, common approaches are applied in the supply of both broilers and layers (Figures 4.8 and 4.9).

Process Stage	Consideration
Elite/great-grandparent/grandparent/parent breeding birds ↓	Health Vaccination Competitive exclusion Biosecurity Air filtration Waste control Feed control Personnel control Hygiene Cleaning and disinfection Testing
Broiler ↓	Biosecurity Waste control Feed control Personnel control Hygiene Cleaning and disinfection Testing
Broiler receipt at processor ↓	Supplier quality assurance programme Transport cage design, hygiene, cleaning and disinfection Environmental hygiene *Salmonella* status of flock (positive flock last)
Shackle, stun, kill and bleed ↓	Environmental and equipment hygiene
Scald ↓	Water temperature/treatment Counter flow of water Cleaning and disinfection

Figure 4.8 Process flow diagram and technical considerations for poultry production.

Process Stage	Consideration
Pluck ↓	Equipment hygiene Cleaning and disinfection
Eviscerate ↓	Hygiene Cleaning and disinfection
Inside/outside washer ↓	Hygiene Water quality Temperature
Chill ↓	Hygiene Temperature
Portioning ↓	Temperature Time Hygiene
Pack ↓	Pack integrity
Retail ↓	Temperature Shelf life
Customer	Storage temperature and shelf life Cooking instructions Food safety tips

Figure 4.8 *Continued*

Broilers and layers are reared from young chicks in enclosed barns on poultry farms. These farms are usually under contract to supply a particular egg producer or broiler processor. The chicks are supplied to the poultry farm by hatcheries that receive fertilised eggs from parent birds, bred at breeder farms. The parent breeder farms receive the grandparent birds from a very small number of grandparent breeder farms. They in turn get their birds from great-grandparent/elite stock, grown only by a few suppliers in Europe. It is the elite stock that is bred with specific genetic traits required by the laying or broiler industry, product yield being a major determinant.

As nearly all birds are derived from a very small number of elite/great-grandparent and grandparent supply points, it is clear that any pathogens entering the population in these early stages will be able to spread throughout the layer and broiler populations very quickly indeed.

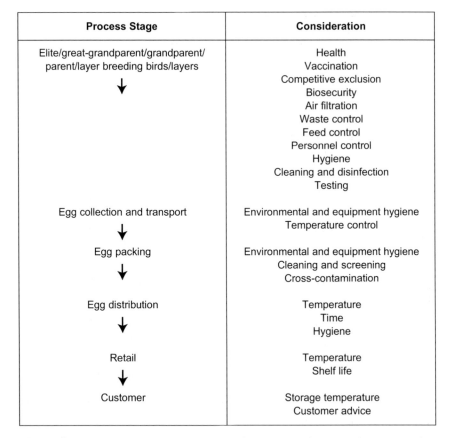

Process Stage	Consideration
Elite/great-grandparent/grandparent/ parent/layer breeding birds/layers ↓	Health Vaccination Competitive exclusion Biosecurity Air filtration Waste control Feed control Personnel control Hygiene Cleaning and disinfection Testing
Egg collection and transport ↓	Environmental and equipment hygiene Temperature control
Egg packing ↓	Environmental and equipment hygiene Cleaning and screening Cross-contamination
Egg distribution ↓	Temperature Time Hygiene
Retail ↓	Temperature Shelf life
Customer	Storage temperature Customer advice

Figure 4.9   Process flow diagram and technical considerations for egg production.

Poultry feed producers, sometimes independent of the egg and poultry production industry, although now increasingly part of integrated businesses, supply feed used for broilers and layers and contamination of feed has also been shown to be a significant source of *Salmonella*.

Poultry

Broilers are reared for 5–8 weeks depending on the size of the bird required and then they are transported in cages on lorries to slaughter and processing facilities (Figure 4.8). Cages are loaded onto automatic conveyors that empty the birds into a collection chamber where they are transferred by hand onto shackles. They then travel upside down on automated lines through the entire process. First, the birds are stunned, usually through the application of an electrical stunner. After stunning they are killed by cutting the blood vessels at the sides of the neck and the birds are bled before moving into a scald tank. The scald tank contains a large volume of water at a temperature of *c.* 50–60°C. The temperature used, soft scalding nearer 50°C and hard scalding nearer to 60°C, depends on the end use of the bird but the aim at this stage is to get all of the feathers completely wet and loosened so that the subsequent de-feathering/plucking stage achieves full removal of all feathers. As the bird has yet to be plucked and eviscerated, the scald tank becomes heavily soiled during the course of production and can be a significant source for cross-contamination of pathogens to birds passing through the tank. The plucking system also facilitates the spread of contamination between birds. Plucking machines consist of a large array of rubber fingers that, when rotated, flay at the birds and pull the feathers out. Clearly, this unit has significant potential to damage the skin of the bird and it is carefully aligned to ensure that maximum feather removal occurs with minimum damage to the epidermal layers of the birds. Nevertheless, because of the design and physical nature of the rubber fingers, they are readily contaminated with microorganisms that are easily spread to other birds passing through the plucker.

After plucking, the birds are conveyed to an evisceration unit that cuts a hole in the caecal vent area through which the viscera is extracted. Other organs are also removed and, depending on the design of the unit, vacuum systems can ensure full removal of organs from the cavity and blades can trim the neck flap. Most systems have washing units that spray water inside and onto the outside of the carcass, referred to as inside/outside washers, after evisceration. Carcasses then pass into a chiller where they are chilled to < 5°C overnight. Alternatively, they may be passed through a chilled water tank and then frozen.

Chickens may be packed and sold whole or they may be portioned into wings, breast and thigh meat before packing. Chicken and chicken portions are usually placed into a plastic foam tray containing a soak pad and over-wrapped in plastic film. They are packed usually without any modified atmosphere and sold with a shelf life of 6–8 days under chilled conditions or several months as a frozen product. Fresh, chilled chicken rapidly succumbs to spoilage mainly by pseudomonads that produce offensive, putrid odours characteristic of spoiled chicken.

Eggs

Eggs are produced by laying hens that are raised principally in two different ways, as battery hens or free-range birds. Battery hens are confined in cages throughout their laying life which can be 12–18 months. Eggs, laid in the cage, usually drop into troughs or are conveyed to a collection bay. Eggs are collected (Figure 4.9) on a daily basis, trayed and boxed and transported to central packing stations where they are examined for soiling, cracks or deterioration (candled) and then graded (class and size) and packed into cardboard or plastic egg boxes. In some countries, eggs are brushed or water washed and/or waxed. Free-range hens are able to move around the environment and are not continuously confined to an enclosed shed. Consequently eggs can be laid on the ground and in a variety of other uncontrolled areas. Due to the nature of free-range conditions, it is more difficult to control insects, pests and wild birds, and flock and egg contamination from such sources can present a problem. Free-range eggs are hand collected on a daily basis and then transported to a central packing station where they are packed and distributed.

Eggs are usually allocated shelf lives of 21–28 days from laying and, in the UK, are not chilled during distribution or retail display. It is normal for egg packers to store and distribute eggs at temperatures below 20°C, the recommended maximum temperature for storage by the Advisory Committee on the Microbiological Safety of Food (1993). Eggs are delivered daily into most major retail outlets within 2–5 days from laying and are sold within a further 1–2 days. Eggs are usually labelled with a recommendation to consumers to refrigerate them after purchase due to the potential for temperature abuse in the home but many people store eggs at ambient temperature throughout the shelf life.

Raw material issues and control

For chicken and egg producers, the principal 'raw material' is the chicken itself. Although the type of chicken used by different farms may be derived from different stock, geared to the end user requirements, the methods of producing the birds are very similar.

Production of chicken starts with the elite/great-grandparent stock (Figure 4.10). In Europe, these are bred at one of only a few centres and it is here that the genetic traits of importance to the ultimate quality and yield of the chicken are developed through active breeding programmes. This elite stock is used to supply the grandparent stocks that are maintained by breeders usually based in the country of use. Again, there are only a limited number of grandparent breeding facilities. These then supply the parent stock of birds to commercial breeder farms or multiplication centres, who in turn supply eggs to hatcheries who supply the chicks to broiler or egg-layer farms.

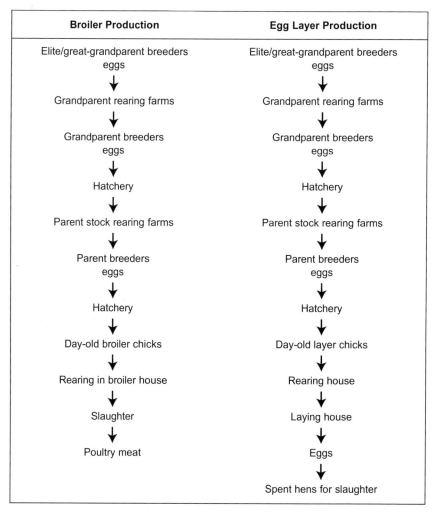

Figure 4.10 Poultry and egg breeding production chains, adapted from Advisory Committee on the Microbiological Safety of Food (1996).

Different levels of biosecurity are required for each of the different stages of production. It is clear that with so few facilities breeding elite and grandparent stock in the world, any birds infected with *Salmonella* at these stages have huge potential to spread these organisms throughout stocks of egg-laying and broiler-producing birds. As a consequence, the level of control operated in these primary production stages must be, and is, exceptionally high. Effective controls must be exerted over the production environment and all feed and personnel entering these sites. It is common for these facilities to be operated totally as high-risk factories with segregation of the hatching and rearing facilities. Any eggs, received on to the site are usually decontaminated using chemical fumigants. Personnel entering such facilities must also undergo a full change of clothing, often including showering, between the low- and high-risk areas of the plant. Most feed is heat treated and it is all subject to rigorous microbiological testing prior to use. In addition, the air supply is often filtered.

Competitive exclusion techniques, where the day-old chick is sprayed with organisms derived from the faecal material of healthy adult birds may be employed at the breeder stages. This has been shown to confer some increased resistance to *Salmonella* infection in the early days of the chicks' development due to the competitive effect of the adult birds' microflora, i.e. they occupy the sites in the gut that could otherwise be colonised by *Salmonella*. Such techniques are more commonly employed for breeder stocks and laying hens, but due to the costs involved they are rarely used for the broilers themselves.

Corrier *et al.* (1998) reported the effect of a competitive exclusion culture on the colonisation by *Salmonella* of broiler chicks. Batches of one hundred newly hatched (day 1), *Salmonella*-free chicks were placed in a cabinet and spray treated with 20 ml of competitive exclusion culture. Three inoculum levels of the competitive exclusion culture were used to deliver 10^6 cfu, 10^7 cfu and 10^8 cfu of culture organisms per chick. The contamination rates and levels in the caecal contents on day 10 were compared to controls after oral challenge with *S.* Typhimurium (1×10^4 cfu) on day 3. The incidence and actual level of caecal colonisation was significantly reduced in chicks administered the competitive exclusion culture. The average rate of *Salmonella* isolation from control (untreated) chicks was 90% with a mean level of log 4.12 cfu/g of caecal contents. The rate of isolation decreased to 53%, 45% and 35% and the mean level of contamination to log 1.82, log 1.38 and log 1.16 cfu/g of caecal contents of chicks administered competitive exclusion cultures at levels of 10^6 cfu, 10^7 cfu and 10^8 cfu organisms, respectively. This level did vary across the triplicate trials but in all cases the protective effect of the competitive exclusion culture against *S.* Typhimurium was significant (Table 4.35).

In a further experiment using day-old chicks administered the 10^7 cfu competitive exclusion culture per chick, the rate, level and spread of contamination was compared to control (untreated) chicks. Competitive exclusion (CE) culture administered chicks and control chicks were kept in separate pens. On day 3, wing bands were attached to half of the CE

Table 4.35 Effect of competitive exclusion culture administered to newly hatched chicks on caecal colonisation by *S.* Typhimurium, adapted from Corrier *et al.* (1998)

Level of competitive exclusion culture administered per chick*†	Percentage incidence of *S.* Typhimurium in caecal contents (range from 3 trials)‡	Level of *S.* Typhimurium in caecal contents, log cfu/g (range from 3 trials)‡
0 (control)	90 (85-95)	4.12+/−2.23 (3.83-4.60)
10^6 cfu	53 (20-80)	1.82+/−2.02 (0.58-2.72)
10^7 cfu	45 (35-55)	1.38+/−1.95 (1.10-1.95)
10^8 cfu	35 (0-55)	1.16+/−1.69 (0-1.78)

* Chicks treated with competitive exclusion culture on day of hatch (day 1).
† Total 60 birds per treatment, caecal contents analysed on day 10.
‡ Chicks orally administered *S.* Typhimurium on day 3.

treated and control chicks and these chicks, the 'seeders', were orally administered *S.* Typhimurium (1×10^4 cfu) and then replaced in pens with the remaining chicks, the 'contacts'. The chicks were reared for up to 43 days. Incidence and levels of *S.* Typhimurium were assessed in caecal contents after 21 days and 43 days and also in litter samples at the same time points (Table 4.36). After 21 days, 88% of caecal contents from 'seeder' control chicks (administered no competitive exclusion culture) were found to be colonised after oral administration of *S.* Typhimurium and 87% of the control 'contacts' were colonised. After 43 days the caecal contents of control 'seeders' were colonised at an incidence of 11% and the 'contacts' at 7%. Administration of 10^7 cfu/g competitive exclusion culture to each newly hatched chick reduced the incidence of caecal colonisation by *S.* Typhimurium to 17% in 'seeders' after 21 days and to 10% in 'contacts'. By day 43, this incidence had reduced to 0% in both 'seeders' and 'contacts'.

Clearly the use of competitive exclusion techniques can help to reduce the incidence of *Salmonella* carriage in chicks, although the efficacy of such treatment will vary according to, amongst other things, the composition of the competitive exclusion culture used, the point at which it is administered, the method of administration and the level of *Salmonella* contamination in the flock. The prospects for competitive exclusion have been reviewed by Mead (2000).

It should be remembered that whilst most strains of *Salmonella* do not cause significant illness in poultry flocks, some strains cause significant morbidity and mortality in the flock. Avian strains such as *S.* Pullorum and

Table 4.36 Effect of treating newly hatched chicks with a competitive exclusion culture on the colonisation of caecal contents and contamination of litter after oral challenge with *S.* Typhimurium, adapted from Corrier *et al.* (1998)

Level of competitive exclusion culture administered to chick (cfu)*	Level of *S.* Typhimurium administered to chick (cfu)†	Age of chick at sampling (days)	Colonisation by *S.* Typhimurium			
			Caecal sample		Litter sample	
			Incidence (%)	Mean level (log cfu/g)	Incidence (%)	Level (log cfu/g)
0	0‡	21	87	3.90+/−1.45	92	2.78+/−0.75
	1 × 10⁴		88	3.84+/−1.62		
10⁷	0‡		10	0.43+/−0.67	47	0.96+/−0.58
	1 × 10⁴		17	0.63+/−0.95		
0	0‡	43	7	0.19+/−0.24	33	0.66+/−0.84
	1 × 10⁴		11	0.32+/−0.35		
10⁷	0‡		0	0	14	0.28+/−0.68
	1 × 10⁴		0	0		

* Administered by spray to newly hatched chicks (day 1).
† Administered orally ('seeders').
‡ No direct administration of *S.* Typhimurium but reared in contact with birds administered 1 × 10⁴ cfu *S.* Typhimurium ('contacts').

S. Gallinarum that cause bacillary white diarrhoea and fowl typhoid respectively can be very serious infections in the flock. In addition, *S.* Enteritidis can be exceptionally invasive and can increase mortality in the flock, particularly in young chicks. Indeed, it is the invasive nature of this strain and its reported ability to infect the reproductive organs that has allowed it to be very successful in its spread through the poultry population as it can spread through the faecal/oral route and through the maternal/egg route.

A vaccine developed against *S.* Enteritidis can also be employed during the breeding process for broilers and layers, but it is mostly applied in the case of egg production, at the layer stage. In the UK, it is now common practice for laying hens to be vaccinated against this organism. The reason why vaccination is employed for laying hens and not usually for broiler hens relates to cost efficiency. Laying and breeding birds are productive for much longer periods than the life of broiler hens, which are reared for up to eight weeks only before slaughter. There are therefore fewer layers and breeders and vaccination costs are spread over a longer timescale. The increased vaccination of laying hens brought about by the UK egg industry code of good manufacturing practice (Lion Code, Anon, 2000g) is believed to have been a significant factor in the reduction of *Salmonella* Enteritidis in flocks and eggs and thus, in the reduction in related infections reported in the human population in the UK.

Vaccines are available for a number of purposes and are of a variety of types. These include both live, attenuated strains of *Salmonella*, most commonly *S.* Enteritidis or *S.* Typhimurium and dead vaccines. Live vaccines are administered in the feed or water and act by colonising the intestine and stimulating a natural immune reaction that provides cross-protection against invading strains of *Salmonella*. Clearly, live vaccines must not be harmful to either the chicken or humans and for diagnostic purposes must also be able to be differentiated from other *Salmonella* strains when examining poultry and poultry related samples for contamination with *Salmonella*. Because they are live, any intervention used for therapeutic purposes as part of animal husbandry, such as the administration of antibiotics, may destroy the vaccine strain and compromise any protection given. Dead vaccines operate by creating immunity in the bird following injection of each individual bird. They can be highly effective and provide a long-lasting level of immunity to the target and closely associated strains. The high labour cost associated with the administration to each bird however does mean that the cost of using such vaccines is greater. It is for this reason that they are generally administered only to laying birds and to breeder birds.

During all breeding stages, hatched chicks and associated material are subject to a high level of microbiological examination, involving sampling of meconium (a material from the alimentary tract) from day-old chicks. Spent shells and dead chicks are also routinely tested together with a wide variety and large number of environmental samples including dust, feathers and litter. Such testing has been a requirement for all breeding flocks under the Poultry Breeding Flocks and Hatcheries (Registration and Testing Order) 1989 (Statutory Instrument Number 1963). Grandparent stock breeders often employ a more extensive sampling programme than that required in law as every effort is made to ensure that contamination by *Salmonella* does not arise in these primary stages of the rearing process and the industry is very well structured and operated to achieve this. It remains clear that it is absolutely vital to ensure that primary (elite and grandparent) stock is maintained free from contamination with *Salmonella*.

In addition to these early stages in the rearing chain, significant potential exists for generating infected birds in the multiplication centres or parent breeding farms. Although it is normal for these growers, on receipt of eggs or chicks, to employ rigorous intake testing programmes for *Salmonella*, these farms are not usually subject to the same level of biosecurity as the grandparent or elite stock units, but nevertheless, they do employ many measures to preclude infection with *Salmonella*. These measures include control of feed supplies and control of personnel entry to the farm, which as a minimum should include the changing of footwear, the use of footbaths containing disinfectant at entrances to the farm, changing of external clothing for overalls or coats dedicated to the area and effective hand-washing/disinfection procedures. The task of these parent farms is to produce a large and continuous supply of fertilised eggs for hatcheries where the eggs are hatched and the chicks supplied to the layer or broiler farms.

The hens and cockerels in parent stocks are for breeding purposes and, like laying hens, are kept for many months, so any contaminants entering the shed have long periods in which to spread and eventually colonise the entire flock. Control must be exercised over the microbiological integrity of the feed, water supply, and litter in addition to the biosecurity of the poultry house/shed to prohibit infestation with rodents and insects. These factors will now be considered in more detail in relation to broiler and layer farms but similar issues exist at all levels of poultry production.

Sweden is perhaps one of the few countries in the world where *Salmonella* is effectively controlled in broiler and layer flocks. Controls to reduce *Salmonella* in Swedish meat production were stimulated by a devastating outbreak of salmonellosis in 1953 which resulted in 8845 cases and 90 deaths attributed to raw meat (Lundbeck *et al.*, 1955). In relation to poultry, a programme spanning some 20 years or more identified key stages of the production process that were targeted for action to control *Salmonella*. The measures taken were supported by strict intervention procedures to ensure the destruction of the organism if found in poultry flocks. In an ongoing programme, grandparent and parent flocks are carefully monitored, feed used is almost always heat treated, all ingoing materials are monitored for contamination and high/low-risk separation procedures are generally in place, e.g. for personnel. Although a rare occurrence, flocks found to be contaminated with *Salmonella* are slaughtered and heat processed. Shed litter from contaminated flocks is either incinerated or subjected to controlled composting to avoid proliferation of the organism in the environment. The controls in place have resulted in making the isolation of *Salmonella* a rare occurrence in broiler and layer flocks with a near zero incidence of *Salmonella* in Swedish poultry meat. The controls used have been extensively reviewed by Wierup *et al.* (1995).

From most commercial parent breeder farms fertilised eggs are supplied on a daily basis to commercial hatcheries. On receipt, they may be fumigated, using gaseous formaldehyde, to decontaminate the eggs. Eggs are hatched and the chicks reared under conditions of biosecurity designed to exclude contamination from people and insect or rodent infestation. It is also usual to filter the air supply into hatcheries.

In addition, hatcheries operate significant sampling and testing programmes of materials and the environment during the production of chicks. This includes testing chick box liners, dead chicks and spent shells. In many cases, hatcheries also use competitive exclusion techniques on the day-old chicks or, particularly in the case of laying hens, vaccinate the chicks against *S.* Enteritidis. It is believed that this vaccination programme together with improvements brought about by the widespread adoption of improved practices laid down in egg industry codes of practice (Lion Code, in the UK) may be one reason for the dramatic reduction in *S.* Enteritidis infections in recent years. Infections in England and Wales due to this organism decreased from 23 000 in 1997 to approximately 8500 in 2000 (Anon, 2000a).

Chicks are supplied to broiler and laying farms where they are reared until suitable for either culling for meat production or laying for egg production.

These farms are often the points where *Salmonella* infection of the flock occurs. This can arise from a variety of sources, all of which must be carefully controlled. Sources include the chicks themselves, the fabric of the shed, the feed, water, air supplies, rodents, insects, bedding and external contaminants brought in by people. In order to control these, each aspect must be carefully considered using a hazard analysis based approach and appropriate controls and monitors implemented.

Chicks and chick control have already been discussed, but another important area for control and a common source of *Salmonella* is the feed supply. Feed mills, that are often not part of the poultry industry, produce feed for poultry and often for other farm animals using the same equipment. The operational quality standards in place at feed mills vary considerably as do the production methods. Poultry feed usually consists of a mixture of raw or processed cereal products, fish or bone meal and vitamin and mineral supplements. Use of mammalian by-products in animal feed is widely banned but feed ingredients derived from other species such as fish and poultry may be practised. It is common for feed ingredients to be subjected to a high heat process, but some feed ingredients, such as wheat, is not heat processed. Unfortunately, even if the feed is heat processed, the standard of segregation of raw materials from heat-processed materials has historically been highly variable from good to completely lacking in some feed mills. Indeed, poor levels of biosecurity also allow significant opportunity for insect and pest infestation in some feed mills. Effective biosecurity measures and segregation between raw and heat-processed materials should be a prerequisite for the supply of feed to the poultry and farm animal production industry. Feed supplies should be routinely sampled and samples examined for the presence of *Salmonella* prior to supply and use.

It is fair to say that whilst feed and feed ingredients have occasionally been found to be contaminated with *Salmonella* (Table 1.7), the serotypes isolated are often more unusual varieties and *S.* Enteritidis and *S.* Typhimurium have rarely been isolated from these sources. Nevertheless the finding of any *Salmonella* serotype in feed should be considered to be unacceptable and measures should be put in place to ensure such contaminants do not arise, as is done by some of the major and more reputable feed mill operators. As part of a pathogen control programme, some feed mill operators mix the feed with proprietary acid mixtures that are believed to deliver some reduction in contaminating microorganisms already present or that may get into feed during handling on the farm. These additions should not be used in an attempt to 'clean up' contamination arising from poorly controlled production practices.

It is surprising and disappointing to observe that simple measures to help exclude microbial contaminants from broiler farms are often not in place. A system to prevent entry of personnel into chicken sheds without changing footwear and outer clothing is probably the simplest way to minimise contamination from these sources, but is often the one most commonly omitted. All those entering chicken sheds should change into clean footwear and coats dedicated to the shed and thoroughly wash their hands before entering. The external environment of the sheds is likely to be frequently contaminated with enteric pathogens from wild animals, birds and rodents and, together with a biosecure shed structure, they can be excluded by the adoption of the simple measures described.

Bedding litter used in sheds is another potential source of contamination. Litter may come from a variety of sources including straw or wood shavings and a particular quality of litter is required as it is important for preventing damage and infection of chicken feet during rearing.

Litter, which is not changed during the growing time of a broiler production batch becomes heavily soiled by faecal material from the birds and ammonia levels can build up significantly. Any contamination entering the shed and infecting one bird will be excreted into the litter and rapidly become transferred to other birds, e.g. feed often falls onto the litter encouraging the natural ground searching behaviour of the birds and, when they search for food in the litter, they will also ingest any microbial contaminants present.

Feed should be delivered to the farm in contained units and feeding systems should be employed that help avoid spillage of feed onto litter. The design of the feed system should also preclude the possibility that the feed can become contaminated by the bird faeces. Open chain feeders can be contaminated through deposition of faeces directly into the feeders thereby distributing contaminants throughout the shed. Careful design of the feeding system, ideally using discrete feeders, is important in the prevention of spread of contamination.

Litter is one of the many types of sample taken to determine the presence and incidence of *Salmonella* in the flock. Although not necessarily a frequent cause of *Salmonella* contamination entering the farm, litter removed from sheds of infected birds has the potential to be highly contaminated with *Salmonella* and other pathogens. Although the build up of ammonia in the litter will have some antibacterial effect, the recycling of chicken litter by dispersal on fields for growing agricultural crops still needs to be managed carefully. Many farmers compost the chicken

litter prior to use, but some apply the material to land with little processing that would result in a significant reduction of contaminating organisms. Not only can this potentially compromise the safety of the
agricultural crop, it is possible that continued recycling of *Salmonella* and
other enteric pathogens in the environment in this way provides greater
opportunity for the organism to be reintroduced into the chicken farm
from the local environment. It is interesting to note that the practice
employed in Sweden, where *Salmonella* contamination of chickens has
been virtually eliminated, is to ensure that the litter from any flock where
Salmonella is found is effectively treated to ensure it does not
contaminate the environment.

Water is another means whereby contamination can spread through the
shed very quickly. Water may be supplied in suspended troughs that run
the length of the shed or through discrete water feeders that can be
lowered from the ceiling. Water is a vehicle through which contaminants
can readily pass from one bird to another. Open troughs become contaminated with chicken litter, and faeces are often deposited directly into
the trough. It is easy, therefore, to see how one contaminated bird can
quickly transfer infection to many other birds in the same environment.
The use of discrete water feeders or nipple feeders and regular decontamination of water supplies are important considerations in trying to avoid
colonisation by *Salmonella* in the chicken shed. Most mains water supplies have some residual chlorine present although well waters and
borehole sources will not. The presence of low levels of residual chlorine
in water may be useful in preventing significant proliferation of *Salmonella* but it is likely that the very low levels of chlorine present will have
no effect once the water becomes heavily soiled in the chicken shed.

A regular sampling programme should be operated covering all areas of
the poultry shed to monitor the incidence of *Salmonella*. Samples of dead
chicks, litter, water and feed should be taken at regular intervals and
examined for *Salmonella* and these results should be used alongside
mortality and yield data to monitor the efficacy of the controls in place at
the farm in relation to this important pathogen.

A typical large poultry shed may hold 20 000 birds or more and at some
point the birds need to be collected for slaughter. Some farmers operate a
process of thinning the flock on one or more occasions, during which a
proportion of smaller birds are taken for slaughter, leaving the others to
grow larger. This allows a higher stocking density of birds initially as the
proportion removed during rearing allows more space for the remaining
birds which will grow to a larger size. However, collection of birds during

thinning involves bringing machinery, that is normally stored and used on other parts of the farm, into the shed to round up the birds into cages. This has a significant potential to introduce contamination into the shed from the surrounding farmyard and whilst this does not generally have a big impact on the colonisation of birds being removed, it can leave contaminants capable of colonising and passing throughout the remaining flock. Clearly the best policy is not to thin but if this is done it must be carefully controlled to prevent the introduction of contaminants.

Another potential source of *Salmonella* for poultry flocks is via rodents and other infestation. Rodents are known to be common carriers of *Salmonella* and it has been speculated that they may have been the original source, some four decades ago, of *S.* Enteritidis for poultry flocks. It is interesting also to note that some countries have used rodenticides containing *S.* Enteritidis to control rats (Anon, 1995c) and it is possible that this may have assisted the spread of the organism into chicken farms, i.e. via rodents that were resistant to the material used.

The integrity of poultry houses with respect to rodent infestation must be a high priority for farmers. Windows, which are usually open for ventilation purposes must be appropriately screened to prevent entry and the walls and door entry points must be completely secure. The smallest gap of 1–2 cm is sufficient to allow entry of small mice.

Of course, the ability to control such factors in the environment of free-range broilers or layers is almost completely lost and, consequently, the exclusion of *Salmonella* from rodent, insect and wild bird sources is even harder to achieve (if not impossible) than for layers or broilers farmed in closed environments.

After a broiler house/shed is emptied of chickens, all of the litter is removed and the house is fully cleaned and fumigated. Modern houses are built of steel, plastic and brick with concrete floors and are somewhat easier to clean and disinfect than the older types of construction. However, many older types of chicken shed with wooden walls and earthen floors are still in use today and cleaning and disinfecting these is much more difficult, if possible at all. Clearly, every effort has to be made to prevent sheds from harbouring *Salmonella* left by previous flocks of poultry, as this can be a major source of infection for the incoming flock. Monitoring the efficacy of disinfection programmes should be a component part of the control strategy for *Salmonella*.

Every effort must be made to prevent contamination of the flock occurring on the farm as once it is contaminated it is very difficult to prevent further spread to successive flocks and subsequent processes.

Similar hazards exist requiring similar controls to be exercised in the care of intensively farmed laying flocks. However, because the birds are captive in cages in battery production systems, some different challenges are presented. *Salmonella* Enteritidis and Typhimurium infection occurring in a laying flock has in the recent past in the UK required destruction of the flock and decontamination of the poultry house. Although this was a requirement in the late 1980s and early 1990s, there is no evidence to indicate that these intervention measures, in isolation, actually resulted in reduced human infections with these serotypes. Destruction orders were subsequently confined to the detection of *S.* Enteritidis only (Advisory Committee on the Microbiological Safety of Food, 1993). In relation to the detection of *S.* Enteritidis and *S.* Typhimurium in layer and broiler breeding flocks, destruction of the flock remains one of a number of control options, including antibiotic intervention (Advisory Committee on the Microbiological Safety of Food, 1996). The extensive sampling requirements for poultry breeding flocks and hatcheries are detailed in the UK Poultry Breeding Flocks and Hatcheries Order, 1993.

As mentioned already, it is much more difficult to prevent contamination with *Salmonella* in free-range flocks whether for layer or broiler purposes. Layers will lay eggs on the ground or in nests and control of microbial contaminants is hardly possible.

It is interesting to note that some countries, most notably Sweden, have, by interventions, reduced levels of *Salmonella* contamination in poultry flocks to < 1%. Indeed, many individual farms in the UK have achieved incidence levels as low as this on occasion but are inconsistent and many also have higher incidence. In Sweden, compulsory government-subsidised destruction of infected flocks together with the implementation of strict control measures for feed and litter and destruction of contaminated litter meant that the cycle of infection/contamination from the environment to the flocks was broken and the incidence of *Salmonella* gradually reduced and the organism is now nearly eliminated. It is not impossible to achieve a similar situation in all flocks but the compulsory destruction of flocks without compensation would drive many businesses out of production. Nevertheless, strict implementation of the measures detailed above could significantly reduce the incidence of *Salmonella* in poultry and eggs and this would probably be reflected in the incidence of salmonellosis in the human population.

Some of the key points that must be controlled to minimise the incidence of *Salmonella* in poultry and eggs are detailed in Table 4.37.

Table 4.37 Key points for control of *Salmonella* on a poultry farm

Key point	Issue	Considerations
Feed	Use of non heat-treated feed can allow introduction of *Salmonella*	Use a suitable acid treatment of heat-treated feed to allow greater stability of the final feed. Subject feed to rigorous microbiological sampling and testing.
Feed distribution to birds	Open systems allow faecal contamination to spread through the unit	Use discrete feeders where possible. All feed systems should be regularly cleaned and disinfected.
Water	Faecal contamination of the water can introduce high levels of microbial contamination and common water feeders can spread contamination	Water must be of potable quality (ideally filtered/chlorinated). Take precautions to prevent faecal contamination of water system. Use discrete/nipple drinkers where possible.
Bedding/litter	Straw/wood shavings can accumulate high levels of faeces and contamination during rearing	Prevent spillage of feed onto bedding to avoid chicks eating food material from the bedding.
People	Uncontrolled entry of personnel can introduce contamination into the shed through hands, shoes and clothes	All personnel should change shoes prior to entering the site and put on clean external clothing such as a coat and hat which are dedicated for farm use. Hands should be washed and overshoes or new shoes and a different coat should be put on prior to entering the shed. Personnel entering a shed should do so through a footbath containing disinfectant.

Table 4.37 *Continued*

Key point	Issue	Considerations
Vermin/pests	Rats, mice, birds and animal pets can introduce *Salmonella* into the shed	All sheds should be built and maintained to prevent entry of pests. Walls should be intact, windows must be screened with fine mesh screens and entrances should be secure with no gaps. Ideally, entrances to the shed should have double doors, one providing access from the external environment to a changing room from which another door allows access to the shed.
Chicks	Contaminated chicks entering the shed will result in extensive spread of contamination	Chicks should be sourced from reputable suppliers and subjected to rigorous controls supplemented with relevant microbiological examination during their production.
Vaccination/ competitive exclusion	Vaccination and/or the use of competitive exclusion cultures can help to prevent infection of chicks with *Salmonella*	Employ the use of vaccination and competitive exclusion in accordance with the manufacturer's recommendations but, when used, they must be in addition to all other controls and *not* as a replacement for good farming practice.
Cleaning and disinfection	The shed floor, walls, overheads and the internal environment can be a significant reservoir of *Salmonella*	The shed and equipment (feeders and drinkers) should be designed to be easily cleaned and disinfected. Stainless steel/brick walled sheds with concrete floors and simple interior design that allows raising/removal of feed and water systems for cleaning/disinfection are preferable. Thorough cleaning and disinfection must be carried out each time the shed is emptied.

Table 4.37 *Continued*

Key point	Issue	Considerations
Bird collection	Intermittent removal of growing chickens (thinning) can introduce *Salmonella* from the machinery/people employed to capture a proportion of the birds	Thinning should either be avoided entirely or kept to a minimum number of times during the production cycle. Equipment used should be cleaned/disinfected prior to entry and personnel should employ suitable footwear/clothing changes and hand-washing procedures prior to entry to the shed.

Broilers are transported to a poultry processing facility for slaughter and processing whilst eggs collected from layers are collected and transported to packing stations.

Transportation methods for broilers provide significant opportunity for spreading faecal contaminants between different birds. Bird crates are usually made from wire mesh and are stacked one on top of another resulting in faeces dropping from birds in cages higher in the stack onto birds below. The unusual environment and movement during transportation increases the voiding of waste by the birds and it is clear how contaminants can be readily spread during this stage. The use of better designs of cages that preclude passage of faeces through gratings and restricting food consumption prior to transportation can help reduce the potential spread of contaminants via faeces.

Once the broilers have been removed from the transport cages, the transport unit itself and all cages must be effectively cleaned and sanitised. This is usually done at the poultry processing site and effective methods for removing faeces and decontaminating the cages are essential for preventing the recycling of pathogens from one flock to the next. The recycling of water and lack of terminal sanitisation of crates in many systems represents a major deficiency in hygiene and poses a significant contamination hazard to birds subsequently introduced into such cages.

Flocks are tested routinely for *Salmonella* 2–3 weeks prior to delivery to the processing unit and it is possible to schedule slaughter and processing so that any contaminated flocks are dealt with at the end of the production shift so as to avoid cross-contamination to uncontaminated flocks.

Process issues and control

Poultry

Once delivered to the processing unit, birds, emptied from cages, are collected in a chamber where they are manually turned upside down and the legs attached to shackles. The shackles are part of a large conveying mechanism that moves the birds sequentially through the processing facility.

The early stages of the process are not generally considered to represent major sources of cross-contamination, although appropriate hygienic precautions such as effective cleaning and sanitisation do need to be undertaken, including the shackles and the blades, the conveyor system and the environment. The birds are first stunned by an electrical stunner to render them unconscious. They then pass through a blade unit that cuts each side of the neck, killing the bird and the blood is drained.

It is the next three stages of the process that probably contribute most to cross-contamination in the poultry processing facility. Firstly, the birds enter a scald tank; this is a large vessel containing water at a temperature between 50°C and 60°C. The temperature is critical as growth of organisms needs to be avoided but also the temperature must not be so high as to damage the skin of the bird. The objective of scalding is to ensure that all of the feathers are completely soaked and loosened. As all birds pass through the scald tanks, and the bird has yet to be eviscerated, the tank fills with deposits from the dead bird and rapidly becomes a microbial 'soup'. Contaminants are readily transferred between birds and although the temperatures should reduce levels of contaminants over long periods, the continued introduction of new birds ensures that the microbial loading is maintained at a high level.

After scalding, the feathers are removed using a plucking machine that consists of long rubber 'finger like' projections that successfully remove the feathers from the bird whilst avoiding excessive damage to the skin. Again, as the bird is heavily contaminated on the surface, the rubber fingers themselves rapidly become contaminated and can pass this contamination to successive birds during plucking. After plucking, the birds are eviscerated and the head removed. A small hole is cut in the caecal vent area and the viscera are removed, all in one piece. Spillage of gut or crop contents can lead to contamination of equipment and cross-contamination to all successive birds.

After this stage other internal organs are removed and the cavity is 'vacuumed' to ensure effective removal of all internal contents. The neck flap is usually trimmed and the bird passes through an inside-outside washer.

Following carcass dressing and washing, the feet are cut off and the bird shackled to another conveyor to carry it into a chiller unit. Birds are chilled to <5°C overnight or, if intended to be frozen, they are passed through a chilled water bath and then frozen. Chilled birds may be portioned automatically through machinery that sequentially removes wings, drumsticks, thighs and then breast with each portion falling into large stainless steel containers as they are removed. Significant opportunity exists for cross-contamination during portioning as the pieces drop into containers on top of each other and clearly there is very little that can be done to prevent cross-contamination at this point. In addition, the use of water baths prior to freezing increases the chances of cross-contamination for frozen birds. This may be one reason why, in surveys, such birds are found to be more heavily contaminated than fresh birds.

Whole chicken or portions are packaged into bags and sealed, if frozen, or if chilled, are usually placed into a plastic foam tray containing a soak pad and then over-wrapped with film.

Once the organism is introduced into a slaughter or processing facility it is very difficult to prevent extensive spread of *Salmonella*. With such highly automated processing lines, contaminants can spread via all of the equipment, particularly those pieces with which all birds come into contact. Although scald tank water is continually replenished because water is lost by retention in the feathers and on the skin of the birds passing through the tank, once contaminated, the level of organisms in the water stays fairly constant (International Commission on Microbiological Specifications for Foods, 1998) and *Salmonella* can survive (particularly at the low scald temperatures) in the organic 'soup' that develops quickly in the tank as soil (from the feet of birds), blood, faeces, etc. build up. If *Salmonella* contaminates the scald tank water then this will rapidly spread to the rest of the downstream equipment. Scald tanks, pluckers and eviscerators are all well known as points where organisms including *Salmonella* can contaminate the system and be spread to subsequent birds.

Carramiñana *et al.* (1997) studied the distribution of *Salmonella* in a Spanish poultry slaughterhouse over the course of 20 different production runs. The incidence of *Salmonella* in composite faecal samples taken from crates in which the live birds were delivered to the slaughterhouse was 30%. Of the samples taken from the neck cutting blades, 50% were contaminated with *Salmonella* and 75% of the scald tank water samples were also contaminated. Incidence on poultry carcasses post de-feathering (plucking) and prior to vent cutting was 55% and this increased slightly to 60% after evisceration. Contamination remained at this level on the final chilled carcasses (Table 4.38). Although the incidence of contamination was 60% in the final carcasses, the average incidence of *Salmonella* in liver samples examined was 80%.

Data such as these clearly illustrates the fact that once introduced into a poultry slaughter and processing line, it is very difficult to prevent widespread distribution of *Salmonella* amongst poultry carcasses. It is for this reason that poultry flocks should be routinely tested before sending to slaughter to identify those that are contaminated with *Salmonella* so that such birds can be scheduled for processing at the end of a production day. In this way, the potential for cross-contamination to birds that are free from contamination is minimised. This indeed, was a recommendation made by the Advisory Committee on the Microbiological Safety of Food (1996) in its report on poultry meat and is a widely adopted practice in the UK poultry production industry.

Even though it is recognised as difficult to prevent the spread of pathogens during a particular poultry processing run, an opportunity exists at the end of each day's production to break the contamination cycle. All equipment should be subjected to a full, deep (dismantling parts as appropriate) cleaning and sanitisation procedure each day. The clear objective must be to remove/destroy all microbial contaminants that may have entered the plant during the production day so that the subsequent day's production does not start spreading contaminants immediately.

It is very easy to be complacent about the cleaning and sanitisation requirements in the operation of a poultry production facility. The handling of raw poultry may lead some people to believe, wrongfully, that cleaning and sanitisation is of little consequence as the product is raw and should be cooked properly by the consumer. However, like any microbial hazard, the more frequently it is present and the higher the level, the greater the opportunity for it to exploit vulnerabilities in subsequent handling and cooking practices. Effective cleaning and sanitisation are as critical in a raw poultry factory for preventing build up and spread of pathogens as they are, for example, in the production of ready-meals. Attention to detail, particularly concerning the more difficult and highly contaminated equipment such as scald tanks, plucking units and

Table 4.38 Incidence of *Salmonella* contamination at various stages in a poultry slaughter line, adapted from Carramiñana *et al.* (1997)

	Environmental sample				Carcass sample				
	Faeces on crates	Knife blades	Scald water	Wash water	Post-plucking	Post-vent cutting	Post-evisceration	Post-spray washer	Post-air chiller
Number positive for *Salmonella*/number tested* (%)	6/20 (30)	10/20 (50)	15/20 (75)	1/2 (50)	11/20 (55)	9/20 (45)	12/20 (60)	7/10 (70)	12/20 (60)
Number positive for *S.* Enteritidis	2	7	11	0	7	8	10	7	10

*Inclusive of *S.* Enteritidis isolations.

eviscerators as well as other product contact equipment is critical. Cleaning and sanitisation operations should be routinely followed by monitoring of their efficacy using indicators of microbial contamination such as coliforms or *E. coli* and/or using rapid alternative methods such as ATP-bioluminescence hygiene monitoring systems.

Due to the difficulties in preventing cross-contamination during poultry processing, a number of researchers have investigated alternative techniques for decontaminating poultry carcasses in an attempt to reduce the levels and incidence of *Salmonella*. Wang *et al.* (1997) demonstrated a 0.7–1.6 log reduction in *S.* Typhimurium inoculated onto chicken skin (38.5 cm^2 cut from carcass and mounted on a plastic holder) and subjected to a water spray at temperatures of 10°C, 35°C and 60°C applied at five different pressures (206.6 kPa–1034.2 kPa). Spraying using the same conditions of temperatures and pressure but with 10% tri-sodium phosphate or 0.1% cetyl pyridinium chloride, followed by rinsing in tap water, resulted in a 1.5–2.3 and 1.5–2.5 log reduction, respectively (Table 4.39).

Using a modified inside-outside washer, Yang *et al.* (1998) compared the efficacy of four chemical sprays for removing *S.* Typhimurium, inoculated at 10^5 cfu/carcass, from poultry. Chicken carcasses were spray inoculated with 1 ml of *S.* Typhimurium culture on the breast and the back side and 1 ml in the cavity and left for 30 minutes at 20°C prior to rinsing with tap

Table 4.39 Reduction of *Salmonella* on poultry skin by washing with a variety of washing chemicals and pressures, adapted from Wang *et al.* (1997)

Treatments		Number of *S.* Typhimurium, log cfu/skin* (log reduction)		
Washing solution	Pressure of spray (kPa)	10°C	35°C	60°C
Control	0	7.7	7.3	7.0
Tap water†	206.8	6.7 (1.0)	6.4 (0.9)	5.9 (1.1)
	1034.2	6.9 (0.8)	6.4 (0.9)	6.3 (0.7)
0.1% cetyl pyridinium chloride‡	206.8	5.8 (1.9)	5.8 (1.5)	4.7 (2.3)
	1034.2	5.7 (2.0)	5.6 (1.7)	5.0 (2.0)
10% tri-sodium phosphate§	206.8	6.2 (1.5)	5.3 (2.0)	5.4 (1.6)
	1034.2	6.1 (1.6)	5.3 (2.0)	4.8 (2.2)

* Mean of two replicates (total of 6 skins).
† Maximum reduction of 1.6 log cfu achieved at 620.5 kPa and 35°C (not shown).
‡ Maximum reduction of 2.5 log cfu achieved at 413.7 kPa and 60°C (not shown).
§ Maximum reduction of 2.3 log cfu achieved at 827.4 kPa and 60°C (not shown).

water to remove loosely bound cells. Chicken carcasses were washed in an inside-outside washer at a pressure of 413 kPa using the following solutions, all at 35°C; 10% tri-sodium phosphate (pH 12.3), 0.5% cetyl pyridinium chloride (pH 7.6), 2% lactic acid (pH 2.2) and 5% sodium bisulphate (pH 1.3). The residence time in the inside-outside washer was 17 seconds, after which the carcasses were removed from the shackle line and left to stand for 60 seconds prior to rinsing with tap water (551 kPa) for 17 seconds. Enumeration of *S.* Typhimurium was carried out using a whole carcass rinse technique.

The water spray alone reduced levels of contamination by up to 0.56 log cfu. The four chemical treatment processes however achieved larger reductions ranging from 1.7 to 2.0 log cfu (Table 4.40).

Table 4.40 Effect of applying four different washing treatments using an inside-outside washer on the reduction of *S.* Typhimurium on chicken carcasses, adapted from Yang *et al.* (1998)

Treatment	Reduction in *S.* Typhimurium versus unwashed control (log cfu per carcass)
Water	0.42
10% tri-sodium phosphate	1.78
Water	0.39
0.5% cetyl pyridinium chloride	2.01
Water	0.56
2% lactic acid	1.77
Water	0.23
5% sodium bisulphate	1.70

Other researchers have examined the effect of varying the scald water temperature (Slavik *et al.*, 1995) and pH (Humphrey *et al.*, 1984) in reducing *Salmonella*. Although some reductions can be achieved, it is not possible to achieve a significant reduction given the short time the birds are in the scald tank, the temperature limits within which the scald tank has to be operated to prevent adverse quality effects on the carcass and that some protection is afforded to organisms by the organic debris that can build up during processing.

Eggs

Contamination sources at egg-packing stations are somewhat less problematic to solve than those on poultry processing lines. Eggs are received and stored in warehouses where the trays of eggs are deboxed then usually automatically sorted into sizes and candled to check for cracks and spoilage prior to packing. The biggest sources of cross-contamination in egg-packing houses are any systems that come into direct contact with the eggs themselves. Therefore, egg brushes (if used) or any cups used for picking the egg up or indeed supporting the egg are some of the key areas where contaminants can be spread to subsequent eggs. In order to prevent entry of any *Salmonella* from the shell surface into the egg, UK produced eggs are kept dry and control of surface dirt and debris is by visual inspection and removal of soiled eggs instead of washing. Disinfectant treatment of eggs does not destroy bacteria that have already penetrated the shell, however, mild heat treatment of whole shell egg to eliminate any *Salmonella* that may be on the shell or within the egg contents is possible (Hou *et al.*, 1996) but conditions of treatment have to be finely controlled to prevent the egg from starting to cook thus changing the raw egg quality for culinary use.

Fluctuations in temperature need to be avoided during storage and transportation due to the potential for such fluctuations to allow condensation to build up on the egg, again aiding the ingress of the micro-organisms, if present, from the outside to the inside of the egg. Storage and transportation temperatures should be maintained below 20°C to reduce the opportunity for any contaminants that have entered the egg to elevate in number during the shelf life of the egg. These and other recommendations relating to safe handling practices for eggs were made by the Advisory Committee on the Microbiological Safety of Food (1993) in its report on eggs.

Final product issues

Poultry

Once packed, fresh chilled poultry is given a shelf life of 6–8 days and frozen poultry 9–12 months. Fresh poultry rapidly succumbs to spoilage by pseudomonads which cause a characteristic putrid odour. This occurs particularly in the cavity of the bird or in areas in contact with the packaging of the chicken where moisture is trapped. Growth of *Salmonella* in fresh poultry is not considered to be likely and therefore of significance as chilled poultry should be stored at < 5°C otherwise it will spoil rapidly. Poultry is not treated with any preservative and any *Salmonella* present survives well on both chilled and frozen product as seen from the results of many surveys of such products.

A variety of surveys have been carried out on the incidence of *Salmonella* in poultry and whilst the incidence reported varies, i.e. from 10% to 70% or greater, a notable exception to these high levels is found in Swedish poultry, where they are < 1%.

Uyttendaele *et al.* (1999) reported 36.5% incidence of *Salmonella* on raw poultry products on sale in Belgium. Raw processed chicken was more commonly contaminated (68.3%) than whole chicken carcasses (29.3%) and *S.* Enteritidis was isolated from only 5.4% of samples. Poultry products with skin on were more frequently contaminated with *Salmonella* (47%) than when the skin was not present (34.6%) (Table 4.41).

Table 4.41   Incidence of *Salmonella* in raw poultry products on sale in Belgium, adapted from Uyttendaele *et al.* (1999)

Product	Number of samples positive for *Salmonella*/number samples tested (%)*
Carcasses	
Chicken	39/133 (29.3)
Boiling hen	13/32 (40.6)
Spring chicken	13/48 (27.1)
Guinea fowl	6/32 (18.8)
Cuts	
Chicken	99/225 (44.0)
Turkey	60/164 (36.6)
Spring chicken	8/28 (28.6)
Guinea fowl	0/3 (0)
Processed products	
Chicken	28/41 (68.3)
Turkey	16/66 (24.2)
Total	282/772 (36.5)

*> 1 cfu/100 cm^2 or 25 g.

Plummer *et al.* (1995) reported on a small survey of poultry products in the UK. They reported an overall incidence of *Salmonella* of 22.8% and the incidence on fresh supermarket purchased birds was lower (18.6%) than supermarket purchased frozen birds (25.5%) or butcher-shop/market stall purchased birds (24.5%) (Table 4.42), although, these results were not significantly different in statistical terms. In common with many other surveys, the samples of giblets yielded higher incidence of *Salmonella* contamination (37.1%). The results for birds themselves was lower than those reported several years earlier by Roberts (1991) who reported an

Table 4.42   Incidence of *Salmonella* in retail raw chicken products in the UK, adapted from Plummer *et al.* (1995)

Product	Number of samples* positive for *Salmonella*/number of samples tested (%)
Whole birds	19/64 (29.7)
Breasts	28/91 (30.8)
Quarters (excluding breast)	12/75 (16)
Drumsticks	5/29 (17.2)
Thighs	1/28 (3.6)
Wings	3/10 (30.0)
Mixed portions	6/24 (25.0)
Mince	0/3 (0)
Total	74/324 (22.8)

* 25 g sample of a mixture of skin and muscle from multiple points.

incidence of *Salmonella* in UK poultry of 48%. In fact over the last two decades the incidence of *Salmonella* in retail poultry in ·the UK has reduced significantly, to 5.8% in 2001, due to continued efforts by the industry to improve control of this organism (www.foodstandards.gov.uk).

Eggs

Surveys for *Salmonella* in eggs reveal a much lower incidence of contamination than poultry carcasses. In a large survey of British and imported raw shell eggs in 1991, a total of 7045 samples of British and 8630 samples of imported eggs were examined for the presence of *Salmonella* (de Louvois, 1993). Each sample consisted of a composite of six eggs and no differentiation was made between contamination of the external surface or contents. *Salmonella* spp. were isolated from 65 (0.9%) samples of British eggs and from 138 (1.6%) samples of imported eggs (Table 4.43). However, *S.* Enteritidis was more commonly isolated from British produced eggs (0.7% British, 0.2% imported). Of the 66 *Salmonella* isolates from the 65 positive British egg samples, 33 were *S.* Enteritidis PT4, 15 were *S.* Enteritidis (other phage types), 8 were *S.* Livingstone, 6 were *S.* Typhimurium, 1 was *S.* Infantis and 3 were unnamed serotypes. The predominant serotypes from the 138 isolates recovered from imported eggs were *S.* Infantis (55 isolates), *S.* Livingstone (31 isolates), *S.* Enteritidis PT4 (16 isolates), *S.* Typhimurium (9 isolates) and *S.* Braenderup (8 isolates).

No difference was noted in the incidence of contamination between free-range and intensively produced eggs. From the results of this survey, the

Table 4.43 Incidence of *Salmonella* in a survey of British and imported eggs in 1991, adapted from de Louvois (1993)

	British*	Imported*
Number positive for *Salmonella* spp.†/number of samples tested (%)	65/7045 (0.9)	138/8630 (1.6)
Number positive for *S*. Enteritidis/ number of samples tested (%)	47/7045 (0.7)	19/8630 (0.2)

*Each sample represents the composite of six shell eggs.
† Includes those positive for *S*. Enteritidis.

estimated incidence of *Salmonella* contamination in eggs was calculated to be one in 650 for British eggs and one in 370 for imported eggs.

A survey of eggs on sale in British retail shops conducted between 1992 and 1993 examined the incidence and levels of *Salmonella* on egg shells and in contents after storage at 21°C for five weeks before examination (de Louvois, 1994). The shell surface of each egg was swabbed and the swabs from each six-egg pack were combined and examined for *Salmonella*. The eggs were then immersed in 70% industrial methylated spirit (IMS) for 3–5 minutes to sterilise the shell, allowed to air dry and then the egg was broken and the contents tested for *Salmonella* both by direct enumeration and enrichment culture techniques. A total of 7730 six-egg samples were tested and *Salmonella* was isolated from 17 (0.2%) samples (Table 4.44). Nine positive samples were from the egg shells and eight were from the egg contents, although none of the six-egg composite shell and contents samples were positive for both. Sixteen isolates were

Table 4.44 Incidence of *Salmonella* in a survey of 7730 British eggs stored at 21°C for five weeks, adapted from de Louvois (1994)

	Sampling site		
	Shells*	Contents*	Total
Number positive (%) for *Salmonella* spp.†	9 (0.12)	8 (0.10)	17 (0.22)
Number with contamination (%) > 10^4 cfu/ml	N/A	4 (0.05)	4 (0.05)
Number positive (%) for *S*. Enteritidis	9 (0.12)	7 (0.09)	16 (0.21)

*Each sample represents the composite of six shell eggs.
† Includes those positive for *S*. Enteritidis.

S. Enteritidis, of which 13 were *S.* Enteritidis PT4. The identity of the seventeenth is not clear. The levels of *Salmonella* in the contents of four samples exceeded 10^4 cfu/ml and in these eggs the isolates were *S.* Enteritidis PT4 (3 samples) and *S.* Enteritidis PT1A (1 sample). It was estimated that after storage at 21°C for five weeks, one in 2900 of the retail eggs would be contaminated with *S.* Enteritidis and that one in 6000 egg contents would be contaminated with this organism (de Louvois, 1994). Based on the four eggs where levels of the organism had increased to $> 10^4$ cfu/ml after five weeks at 21°C, it was estimated that such contamination could occur at a rate of one in every 12 000 eggs. This study demonstrates the high numbers that the organism can reach given extended ambient storage conditions and also the ability of the organism to survive for extremely long periods on the shell of the egg.

In a survey of 5790 eggs from 15 flocks naturally infected with *S.* Enteritidis, Humphrey *et al.* (1991) found that the contents of 32 eggs (0.55%) were positive for the organism. The majority of these positive samples had low levels of contamination (< 20 cfu/egg) but five, that had been stored for > 21 days at room temperature had > 100 cfu/egg. Two of these were contaminated with *Salmonella* at levels of 1.5×10^4 and 1.2×10^5 cfu/egg. Free-range eggs from naturally infected flocks were found to be contaminated as frequently as battery laid eggs (0.64% and 0.73%, respectively). It was reported that the albumen was more frequently contaminated with *Salmonella* than the yolk and that storage at room temperature had no effect on the incidence of contamination. However, storage for more than 21 days was statistically more likely to result in high levels of contamination (Table 4.45). Where contamination of the shells was compared with the contents, the shells were more frequently contaminated (1.1%) than the contents (0.9%). In the UK, research of this nature has resulted in recommendations from the Advisory Committee on the Microbiological Safety of Food (1993) to consume eggs within a maximum of three weeks from laying. As a consequence, the majority of UK eggs are individually date marked with a use by date of three weeks from the date of laying.

Results from more recent surveys of eggs demonstrate a similar or lower incidence to that reported from previous surveys. In a survey of 2090 packs of six eggs on sale in Northern Ireland, nine isolates of *Salmonella* (0.43%) were recovered from separate packs of eggs (Wilson *et al.*, 1998). One of the egg packs was contaminated with *Salmonella* in the egg contents and of the nine isolates, three were *S.* Enteritidis (Table 4.46). Eggs from small shops were more likely to be contaminated than eggs purchased from other outlets (Wilson *et al.*, 1998).

Table 4.45 Contamination of eggs with *S*. Enteritidis from naturally infected flocks after storage for different periods at room temperature, adapted from Humphrey *et al.* (1991)

Storage time (days)	Number of eggs sampled	Number positive for *S*. Enteritidis (%)	Level of contamination (cfu/egg*)			
			< 20	< 100	> 100	> 1000
0–7	1085	5 (0.5)	5	0	0	0
8–14	1353	7 (0.5)	7	0	0	0
15–21	1221	1 (0.1)	1	0	0	0
> 21	1603	12 (0.8)	7	0	3	2

*Contents only.

To gain an indication of the level of *Salmonella* contamination in USA laying hens and eggs, a survey of unpasteurised liquid egg and spent hens at slaughter was carried out in 1995 by Hogue *et al.* (1997). Three hundred caecal samples were taken from each individual flock received at a spent hen slaughterhouse (spent hens are laying hens who have come to the end of their laying life) and pooled samples (5 whole caeca per sample) were tested for *Salmonella*. Samples of unpasteurised liquid whole egg samples were taken from egg breaker plants and 10 ml aliquots were tested for *Salmonella* over the same period. A total of 937 samples of unpasteurised liquid whole egg were collected from 20 plants and 17 961

Table 4.46 Incidence of *Salmonella* in a survey of 2090 packs of six raw eggs on sale in Northern Ireland between 1996 and 1997, adapted from Wilson *et al.* (1998)

	Sampling site		
	Shells	Contents	Total
Number positive for *Salmonella* spp.*	8	1	9
Number positive for *S*. Enteritidis	2 (PT4)	1 (PT1)	3
Other serotypes detected	6 (*S*. Mbandaka, *S*. Montevideo, *S*. Typhimurium DT104, *S*. Infantis (× 2), *S*. Kentucky)	0	6

*Includes those positive for *S*. Enteritidis and other serotypes.

pooled caecal samples were collected from 305 flocks. A total of 451 samples (48%) of unpasteurised liquid whole egg were found to contain *Salmonella*. *Salmonella* Enteritidis was detected in 179 (19%) of these samples. Of the 17 961 pooled caeca samples, 7206 (40%) were found to be contaminated with *Salmonella* although from assessments of the contamination level in flocks prior to slaughter, 298 of the 305 flocks (98%) yielded at least one *Salmonella* positive sample over the survey sampling period. A survey conducted four years previously found very similar levels of contamination in unpasteurised liquid egg (53%) and spent hens (24% in pooled caeca samples and 86% of flocks positive) (Hogue *et al.*, 1997). In the case of raw, liquid egg samples, it may be that poorer quality/damaged eggs are preferentially selected for submission to breaker plants as these would be intended for pasteurisation, and this could be reflected in the high incidence of *Salmonella* reported.

General

Recommendations on measures to improve the hygiene of poultry processing were published by a UK government/industry working group on meat hygiene (Anon, 1997e). The following broad points were recommended:

(1) *Poultry flocks:* Measures should be put in place to ensure the health of the flock and where there is evidence of pathogen contamination of a flock, additional systems and measures to minimise cross-contamination should be applied at the slaughterhouse.

(2) *HACCP:* The application of HACCP to the slaughtering process in the production of poultry products was recommended. A framework is given by the group as a starting point for the development of a HACCP system for poultry processing.

(3) Salmonella *testing:* Flocks should be tested 2–3 weeks before slaughter to enable known contaminated flocks to be slaughtered at the end of the production day and after those flocks that have no signs of *Salmonella* infection.

(4) *Process treatments:* The group endorsed the use of processes or treatments to reduce contamination of the final product provided there is a clear benefit to public health and that such treatments are not put in place as an alternative to good hygienic practice.

(5) *Training:* The group recommended the effective training of slaughterhouse personnel and described a set of training targets.

Because of the considerable publicity given to *Salmonella* in relation to poultry and eggs and food safety over the years, and the initiatives taken by governments, the food industry and retailers to provide relevant food safety information to consumers, raw chicken is widely understood by consumers to require special handling. However, the high incidence of gastrointestinal illness still caused by *Salmonella* and *Campylobacter* and involving poultry and egg products, serve to reinforce the fact that problems do remain in the understanding and management of hazards in such products.

It is normal for all raw chicken products to be labelled with cooking instructions giving guidance on the oven temperatures and times required to achieve an effective cook. These are often enhanced by the addition of visual descriptors to provide consumers with simple means of assessing cooking efficacy. Piercing the thickest part of the thigh or breast and checking that the juices run clear is just one method recommended/ suggested in information on the packaging of raw poultry to assist the consumer in ensuring the product is properly and safely cooked. It is also increasingly common to see additional food safety tips on the packaging of such raw foods to reinforce advice regarding the safe storage and handling practices for these products.

In the case of eggs, a number of outbreaks and incidents of salmonellosis are still caused by the consumption of raw eggs or uncooked dishes containing raw eggs. Whilst the incidence of *Salmonella* in eggs is recognised as being very low, the huge number of eggs consumed on a daily basis means that the risk of illness due to the consumption of eggs, particularly those dishes containing raw or lightly cooked eggs where the organism may be given the opportunity to increase in number, is still significant and raw egg products are a particular food safety hazard to the consumer. It is therefore appropriate for advice to be provided to consumers regarding the risks associated with the consumption of raw egg products and advice on the safe consumption of eggs is given by governments and in food safety leaflets generated by trade organisations, food retailers and manufacturers. In some cases, such advice may also be printed on the egg box label itself.

In the UK, following the major salmonellosis 'scares' in the late 1980s implicating consumption of raw eggs, the Department of Health issued advice that the public should not eat 'raw eggs or uncooked foods made from them such as home-made mayonnaise, mousse or ice-cream' (Anon, 1988b). In addition, 'although the risk of harm to any healthy individual from consuming a single raw or partially-cooked egg is small, it is advisable for vulnerable people such as the elderly, the sick, babies and pregnant women to consume only eggs which have been cooked until the white and yolk are solid'. This advice was more recently re-iterated by the UK Department of Health (Anon, 1998e), although the recent drop in numbers of cases of salmonellosis, most probably as a result of the reduction in incidence of *Salmonella* in eggs and poultry, may prompt the government to reconsider whether such advice is still entirely necessary.

The use of raw eggs in uncooked food dishes advocated in recipe leaflets and cookery books and by chefs or 'TV' cooks has been criticised by many observers as promoting unsafe practices. However, it is evident that risks associated with this practice are understood by many chefs who, whilst still choosing to use raw eggs in some dishes, are prepared to communicate the associated risks to their customers/audience (Anon, 1997f). This is perhaps a reasonable approach as it then allows each individual to make an informed choice about the risks they wish to take in what they eat and in the preparation of their own food.

However, the continued preparation and serving of raw egg dishes by many small caterers who neither utilise proper storage conditions or hygiene precautions in their handling and use of eggs nor advise their customers of the potential risks of such dishes, remains a significant concern (Sin *et al.*, 2000).

There is no doubt that *Salmonella*, eggs and poultry will remain inextricably linked for many years to come. However, it is also very clear that the controls introduced by the broiler and egg-producing industries in recent years have resulted in a significant reduction in the incidence of *Salmonella* in these products. As seen in the Swedish example, it is possible, by the continued operation of effective hygiene practices and strict intervention measures, to bring this organism under control. Indeed, it is apparent that this may now be achieved in the UK with eggs, following the introduction of the British egg industry code of practice which advocates measures for the hygienic production of eggs from vaccinated flocks.

GENERIC CONTROL OF *SALMONELLA*

Raw material identified as a potential hazard

A large number of products are made from raw materials that could be exposed to sources of *Salmonella* contamination and that receive little processing capable of reducing it to an acceptable level. In such cases, the control of the raw material is critical to the safety of the final product.

Many of the raw materials for such products are purchased from suppliers and are therefore not under the direct production control of the end product manufacturer. Milk used for making unpasteurised milk soft cheeses, meat for salami, herbs and spices for post-process addition to prepared foods, fresh whole fruit for fruit juicing, etc. may all be 'bought-in' raw materials. The importance of safety in the production of these raw materials cannot be over-emphasised and for this reason they must be included in a comprehensive supplier quality assurance programme. This programme should be designed to ensure that adequate standards of microbiological safety are employed by the raw material supplier. The operation of HACCP-based approaches to food safety by the raw material supplier should be an absolute minimum requirement and this should be routinely monitored through verification checks on batches of materials received and supplier audits. The raw material should ideally be traceable to source and the standards of hygienic manufacture assured at all stages from this point to the point of receipt by the purchaser. Clearly, this is not always possible, as in the case of many herbs and spices and seeds used for sprouting. However, buying from large reputable suppliers often brings with it enhanced traceability and safety assurance programmes as they often contract specific farms for supplies. Some key elements necessary in a supplier quality assurance programme are detailed below.

1. Detailed understanding by the raw material supplier and user of the production process of the raw material and knowledge of the critical control points or those stages in the process influencing the control of Salmonella

In order to effectively influence the safety of a raw material, it is essential that the purchaser/user has some knowledge regarding the production/manufacturing process and associated hazards of the raw material in question. A simple ingredient list (if applicable) and process flow diagram with documented procedures, temperatures and times (as applicable) at each stage will provide an excellent basis on which to assess the micro-biological safety of a raw material. In addition, the purchaser should expect to be provided with a documented HACCP schedule that identifies key hazards and associated controls. This can also be very useful in subsequent auditing of the raw material supplier as focus can be placed on the adherence to critical limits defined in the schedule. The inability to supply such information may be indicative of the absence of effective controls in the processing of the raw materials. Whilst the full implementation of HACCP-based systems is less than complete in the primary agricultural sector, whether or not a raw material supplier has adopted such a system can be a very useful discriminatory tool between prospective suppliers of the same raw material.

2.   *Audit of the raw material supplier to review process control*

Raw material suppliers of ingredients that are critical to the safety of the final product, such as raw meat for salami and milk for unpasteurised milk soft cheeses, should be subject to routine and regular assessment of their compliance with good manufacturing practice and control of specified critical process points. This requires the implementation of a supplier quality assurance programme that includes formal supplier audits. Raw material suppliers in this category should be audited at least annually and, whilst the audit may be conducted by the purchaser or indeed, a third party, the structure and scope of the audit must encompass all points likely to affect the safety of the materials supplied. Many third party audit schemes are now in place to do just this, e.g. schemes auditing to the British Retail Consortium (BRC) technical standard for companies producing retailer own-brand goods. Audits carried out to industry-recognised standards by an accredited third party means that through one external formal auditing scheme, a manufacturer can provide evidence regarding the effective manufacturing controls in place to all relevant customers. In some cases this can significantly reduce the audit burden placed by multiple customers on that supplier.

3.   *Raw material verification checks*

Microbiological examination of incoming raw food ingredients must never be relied upon to ensure the safety of these materials. Statistically, the chances of capturing, in a single small (25–50 g) sample, an organism like *Salmonella* (which is approximately 0.005 mm × 0.002 mm in size) that may be present in only a small amount of material within a batch of several thousand litres or kilograms of material is obviously exceptionally low. This does not mean that microbiological testing has no place in raw material quality assurance programmes, quite the contrary, microbiological testing, particularly when using carefully chosen indicator microorganisms can be a useful way to monitor the hygienic status of a product, but its limitations must be recognised.

Frequent microbiological sampling of raw materials at the point of receipt and testing samples for levels of coliforms/Enterobacteriaceae and *E. coli* can provide useful data for trend analysis by which to assess supplier performance. Supplemented with sampling and testing programmes for *Salmonella* itself, such testing can be a useful component of a raw material quality assurance programme and certainly provide additional assurance that the raw material production process is not out of control.

It is also quite common practice to employ incentive payment schemes in the purchase of some raw materials. This occurs extensively in the purchase of raw milk for the production of unpasteurised milk cheeses. These schemes operate on the principle that providing a low level, e.g. < 10/ml, of an indicator organism such as *E. coli* is consistently achieved in frequent (weekly) tests of farm milk, the farmer receives a premium payment for the milk. On the other hand, the agreement can be structured so that the milk supplier is penalised with a lower payment if levels fall into a higher band, e.g. > 10/ml but < 100/ml. If levels exceed the upper limit, e.g. > 100/ml, then the milk supply may be suspended until levels are brought back under control. Whilst such systems cannot guarantee the absence of pathogenic microorganisms, as they employ microbiological testing only for the presence of indicator organisms, they can encourage the adoption of better hygienic conditions in milk production and thereby reduce the chances of enteric pathogens in particular, entering the raw milk supply.

Extension of this type of incentive payment scheme to other critical raw ingredient supplies such as raw meat, may well be worthy of consideration.

4. Agreed specification with the raw material supplier

It is essential that both parties in the supply/purchase agreement have a clear understanding of the standards that are expected in the supply of the raw material. This is usually detailed in the raw material specification and both parties must understand the importance of all elements detailed in the specification. The high level of microbiological quality required for critical raw ingredients must be clearly understood by the supplier as they may be supplying similar raw materials for a variety of purposes. For example, a farm may be supplying milk to processors who may be subjecting it to pasteurisation prior to manufacturing cheese and also to sites where it is used for the manufacture of raw milk cheeses. The greater risk associated with the latter products may necessitate the employment of specific hygienic practices that must be clearly agreed and documented with the supplier before commencement of supply.

5.   *Conditions of storage and use of the raw material*

Having established effective means of assuring raw material microbiological safety and quality, it is then incumbent on the receiving manufacturer to store the product under conditions that preclude the entry of *Salmonella* or prevent it from increasing in number, if present. Raw material storage facilities should be free from pests and, if refrigerated, should be stored at 5°C or less to ensure complete control of growth of *Salmonella*, although temperatures below 8°C will control the growth of most strains.

The storage location and conditions for different raw materials need also to be considered. Materials to be cooked or subjected to a further process capable of reducing *Salmonella* to an acceptable level should not be stored with those that may be added or used in the final product without any further processing. Care should also be taken to ensure adequate separation of air systems and personnel that could also act as vectors of cross-contamination between these types of material.

Production incorporates processes to reduce the level of *Salmonella* or destroy the organism

A large number of processes employ a stage that is capable of destroying Salmonella or reducing it to an acceptable level either in isolation or in conjunction with other stages. Clearly these stages are critical to the safety of the product and, as such, they must be designed and controlled in a manner that ensures the effective and consistent delivery of the pathogen reduction/destruction stage. Examples include the following:

- heat processing for cooked ready-meals, cooked meats, etc.
- pasteurisation of milk for drinking, for making cheese, yoghurt and other dairy desserts
- chlorination/disinfection of fruit for juicing or vegetables for ready-to-eat prepared vegetables and salads
- fermentation and drying for salamis and raw, dry-cured meats
- fermentation and maturation for hard cheeses

Any critical process stage should be properly validated, using challenge tests where applicable, to ensure that the process parameters necessary to achieve the required reduction are properly understood and verified. The minimum or maximum tolerances that dictate whether the product has received the appropriate treatment must be established and then suitable controls and checks implemented to ensure that the correct process is consistently delivered on each occasion. HACCP-based principles can be applied to process validation.

Cooking

In order to destroy *Salmonella* in cooked meat processes, the minimum commercial processes applied in the UK achieve 70°C for two minutes or an equivalent heat process. To assure safety, all parts of the product must reach this temperature for the specified time. It is, therefore, essential that cooked meat manufacturers establish the conditions necessary to achieve these parameters consistently with their specific processing equipment. This is achieved by means of a process validation study. All such studies must take account of all worst-case conditions that may be encountered and determine what minimum process will consistently achieve the required temperature and time in the product. For a typical cooked meat product, a process validation study should be conducted on meat of the largest possible ingoing size/weight, the lowest ingoing temperature, placed in the coolest part of the oven and with a full oven load. This requires an oven temperature distribution study, ingoing product weight/ size checks and temperature checks.

The product and oven temperatures should be monitored throughout the process, following which, the minimum safe process can be established. The specific factors identified as being essential for safety, e.g. piece size, ingoing temperature, oven temperature and process time, can then be used as in-line process controls and appropriate critical limits specified. Process control monitors should be automated and relevant equipment linked wherever possible to alarm systems so that if the required parameters are not achieved due to a failure in the process, this is immediately registered and appropriate action taken. For example, the process temperature recording systems in all cooked meat operations should be suitably alarmed. Where reliance is placed on steam for heating, then steam pressure should be monitored and linked to an alarm. In addition, if the product is cooked under a grill or in a fryer where it passes under or through the heat source on a conveyor, the belt speed must be assessed as part of the validation study to determine critical speed limits, suitable controls and monitors. In-line process control monitors should be automatically recorded, e.g. using thermographs or continuous printout recorders, and these should be routinely checked and signed off by staff independent of the manufacturing function, e.g. quality assurance personnel. In addition to the in-line process controls, additional manual checks should be made of the product after cooking to ensure it has achieved the target temperature. This is most often done using temperature probes on products exiting the oven, probing products on the top, middle and bottom trays of each rack cooked. It is essential that the target and critical limits associated with these checks are clearly identified as part of the process validation study as an exit temperature of 80°C may be the required temperature necessary to provide assurance that a product has received a process throughout equivalent to a 70°C for 2-minute cook.

Process validation should be carried out for different product types and repeated annually and whenever a new product is introduced or when a process change or new equipment is introduced that may affect the validity of using the results from the previous validation study in the new situation.

A properly validated cooking process and the application of the principles of HACCP should never fail to deliver the heat process required. In order to ensure processes are never unsafe, it is normal to establish target levels and critical levels for process controls that allow corrective action to be triggered, should product not achieve the target level, well before it fails to achieve the critical limit that is the minimum for product safety. In this way the significant costs associated with process failure can be avoided.

In examining reasons for process failures it is not uncommon to find that production pressures brought on by increased demand for a product have led to processors cutting short those stages that are critical to the safety of the product. Reducing cook times, extending production runs and reducing cleaning frequency are all measures that manufacturers may consider taking in order to meet increased volume demands and tight delivery schedules. Production volumes must be considered as part of the hazard analysis of the operation and product safety must never be compromised at the expense of achieving production volumes and supply.

In the case of pasteurised liquid products such as milk, similar validation studies should be carried out and appropriate process controls implemented. Pasteurisers are usually operated to achieve at least 71.7°C for 15 seconds for all the milk and systems are in place to continuously monitor the temperature of the milk in the holding tube. The process temperature probe is linked to an automatic diversion valve that diverts the milk away from the forward flow and into a holding or re-circulation tank if the minimum temperature is not achieved. The temperature at which diversion is triggered is usually set at a temperature above the critical limit, e.g. 72°C to ensure a margin of safety in relation to small temperature fluctuations. Temperature should be continuously recorded and thermograph records should be regularly checked. Diversion valve operation should also be checked prior to starting production to ensure it is correctly set and operating. The variable factor that influences the time that the milk is held at the specified temperature is the flow rate and this is regulated by the process pumps. It is essential that these pumps are maintained and calibrated by qualified personnel to ensure that the flow rate allows the appropriate residence time in the holding tube.

Washing

Some production processes are employed in which the microbial reduction achieved is significantly less than that obtained using cooking processes. The washing of fruit and salad vegetables in the preparation of fruit juices or prepared salads/vegetables respectively, is a good example. Whilst the pathogen reduction achieved is recognised as being low, it nevertheless can still be a significant component of the procedures in place to assure safety of the final product. As such, washing processes should be conducted in a controlled manner that maximises microbial reduction whilst avoiding the spread of contamination further throughout the batch.

The critical elements requiring control are, the quality of the water, the level of the disinfection compound, e.g. chlorine, contact time, and the method of washing. All of these factors must be clearly established and implemented with suitable controls to ensure they are maintained consistently. Use of too low levels of chlorine in common bulk washing tanks may result in ineffective concentrations reaching the microorganisms. Indeed, localised contaminants may simply become washed from one fruit or vegetable and pass, via the water, to the whole batch. Effectively agitated washing systems such as flumes or Jacuzzis are likely to result in better removal of dirt and organisms because of the turbulence they create. However, whilst brush washers may also help dislodge dirt, they are very difficult to clean and can become foci of contamination that, if not properly removed, will then spread to successive batches.

In general, washing in free flowing clean water will probably remove approximately one order of magnitude of microorganisms from the surface of fruit and vegetables. Using chlorine or other disinfectants may increase this to 2–3 log reduction at best. It should always be remembered however that such washing systems have limitations as organisms can become trapped in areas inaccessible to the effects of washing, such as between the leaves of onions, crevices on lettuce and stomata. They can even become internalised in fruits and vegetables in scar tissue and, therefore, washing will only help to remove those organisms directly exposed to the washing process and associated solutions.

Fermentation/drying

Products such as fermented and dried raw meats and many cheeses made from unpasteurised milk rely on the fermentation and subsequent drying process to produce conditions that will effect a reduction in levels of contaminating pathogens that may be present in the raw materials. The pathogen reduction achieved is linked closely to the production, by starter cultures or natural lactic microflora, of organic acids with a concomitant reduction in pH and the adverse effect these have, together with low moisture availability (water activity), on the survival of pathogens such as *Salmonella*.

It is therefore critical to the safety of these products that the temperatures and times of fermentation and drying processes are under effective control. Humidity control of the environment is also important during drying. These parameters in themselves do not affect the microorganisms but they do affect the conditions in the fermented and dried product, which in turn affect the organisms. Therefore, whilst it is essential to understand the process conditions that affect product safety, e.g. temperature/time and humidity, it is important to carry out checks to monitor that appropriate critical changes are occurring in the actual product. Checking of acidity production or pH reduction is often used to ensure that starter culture activity is effectively producing an inhibitory environment for any pathogens. Such changes should be judged against a standard pH/acidity profile of the product obtained during validation studies to ensure the correct profile is achieved consistently. Insufficient or slow development of acidity may not only allow greater survival of *Salmonella* in the subsequent product but it may also permit significant growth of contaminating pathogens during the fermentation stage.

Due to the highly variable nature of many fermentation processes, it is advised that the production processes for salami, raw, dry-cured meats and raw milk cheeses should be challenge tested in appropriate research facilities to establish the capacity of the process and product to deliver a significant reduction in the organism, in this case *Salmonella*.

No matter what the controlling factor or combination of factors in a process, e.g. pH, water activity, heat or washing, it is essential that there is a clear understanding of the effect of each factor or combination of factors on the organism of concern. Suitable controls can then be implemented and monitored to ensure that reduction is reliably and reproducibly effected in routine production processes. The position of the salami or raw, dry-cured meat in the process chamber may influence its rate of fermentation or drying and the position of a meat piece in an oven may influence the achievement of the desired heat process. All such factors must be taken into account in the design and validation of a safe process before any production ever begins.

Product could be re-contaminated with *Salmonella* as a post-processing contaminant

A large number of outbreaks and incidents of salmonellosis implicate cross-contamination of *Salmonella* to ready-to-eat products as a significant causal factor. The principle routes by which the organism gains access to such products are from personnel handling practices, from the inadequate segregation of raw foods from ready-to-eat products and from insects, rodents or other pests. Clearly, if the organism gains access to a high-risk food product environment, then the cleaning and sanitisation procedures and practices in the area become critical in ensuring that spread of the organism is limited and it is effectively removed or destroyed.

Raw/cooked material separation

Any food manufacturing process involving handling of raw materials that may be contaminated with *Salmonella* and incorporating a process step that eliminates or reduces the hazard to an acceptable level, must have sufficient controls in place to ensure the processed product does not come into contact with the raw materials after processing.

Cross-contamination may occur by direct contact between the raw food and the processed/ready-to-eat food or, more often than not, through a secondary vehicle, for example, commonly used surfaces and utensils or people who handle the raw material and then the processed product without taking effective hygienic precautions. Systems must be in place to prevent such simple errors in practice from compromising the safety of the processed product. Processed/ready-to-eat foods are best protected from direct cross-contamination from raw materials by handling them in segregated areas. People, utensils and surfaces must not come into contact with raw foods and then ready-to-eat foods without suitable cleaning and disinfection procedures in place. In most modern food factories, direct or indirect cross-contamination of processed materials is avoided by the operation of a high/low-risk segregation system. Raw ingredients are received, handled and mixed on the low-risk side of the factory by operatives who are dedicated to this side. The processed material/product is then handled and packed on the high-risk side of the factory, again by operatives who only work in this area. The low- and high-risk areas are completely separated from each other by integral walls and the process reduction/elimination stage is built into the partition. For example, a cooker with a double-door entry system may be built across the wall between the two sides. Raw meat is placed in the oven on the low-risk side, the door closed and the product cooked. It is then opened on the high-risk side and removed for further processing and packing. The doors are designed so that they cannot be opened on the low- and high-risk sides at the same time and in this way, providing the cook has been carried out effectively, there should be no opportunity for raw product to come into contact with processed product. The same principle should apply to washing flumes and tanks for decontamination of fruit and vegetables.

Care should be taken to ensure that procedures are in place to avoid contaminated ingredients or utensils entering the high-risk side of the factory. Systems for moving racks, storage bins and other equipment into the area should involve transfer, under controlled conditions, through decontaminating processes such as the cooker or thorough cleaning and chemical disinfection procedures. Ingredients that are added to finished products as post-process garnishes should be sanitised themselves, if possible, or in their containers, to ensure they are not vectors of pathogen contamination. Although principally designed for the control of *Listeria monocytogenes,* high-risk areas in factories are usually kept under positive pressure using filtered air supplies that prevent airflow carrying contaminants from the low-risk preparation areas to the high-risk side. This is particularly important in those factories where the levels of *Salmonella* may be very high in the raw materials, as in the case of raw chicken and meat. Care should also be taken with the design and construction of drainage and waste flows on the high-risk side of factories to ensure that contaminated water cannot flow from the low-risk side through the high-risk area as this may carry significant levels of microbial contaminants including *Salmonella.*

Clearly, this level of control is not possible in most retail or catering facilities. However, similar procedures for avoiding contaminated raw materials or any item in contact with such raw materials coming into contact with the prepared, ready-to-eat food must be in place. Raw and cooked foods should be stored in separate display cabinets or effectively segregated by permanent dividers that are readily cleaned and disinfected. Slicing machines, knives, serving utensils, boards and storage containers used for raw foods must be separate from those used for ready-to-eat foods. Standards of personal hygiene must also preclude the handling of raw and ready-to-eat foods without effective hand-washing procedures and, as employed in most cases, by the use of separate serving utensils.

Personnel handling product

Many ready-to-eat foods are extensively handled as part of the manu-facturing/preparation process. As such, they are at significant risk of becoming contaminated with microorganisms. In most cases, the con-taminants are usually only of significance in relation to spoilage and lim-iting shelf life. However, the potential always exists for an individual to contaminate ready-to-eat foods with pathogenic microorganisms and several outbreaks in the past bear testimony to this route of contamina-tion.

It is imperative that those handling foods are adequately trained and fully aware of the appropriate standards necessary to avoid microbial con-tamination from themselves to the product. A well-trained individual can make an informed decision about practices that are not detailed in work instructions. Ill-informed adherence to instructions may allow improper practices to develop and continue until identified by trained individuals or after a problem occurs, which is often too late.

A prerequisite for the control of enteric pathogen transfer from individuals to foods is the employment of an effective and well-understood infectious disease policy. This should take the form of a medical questionnaire that all staff and visitors must be required to complete for assessment by a qualified person. Questions should be structured to cover situations where the person may be ill themselves with vomiting and/or diarrhoea or, indeed, where others in their family may be ill. In addition, questions should address situations where individuals take holidays in areas where certain enteric diseases are more common or even endemic. Indeed, irrespective of current infection, some employers operate staff stool testing and screening programmes, either as part of employment condi-tions, or on a regular basis. Such policies clearly should be at the discretion of the individual company and it should be recognised that such testing programmes do not, in themselves, guarantee that pathogenic micro-organisms are not being shed intermittently by individuals. It is probably better to assume that all individuals are occasional carriers of *Salmonella* and ensure that adequate hygiene procedures are in place to prevent any such contaminants getting from the intestinal tract onto food. Indeed, in modern societies this should not be too difficult to achieve. Attention should be focused on those occasions, as described above, where the organism may be present and being shed in higher numbers, as in the case of someone with diarrhoea. Some guidance on questions to ask and action to take in these situations is given in Table 4.47.

The primary cross-contamination issue in relation to personnel arises from the hands, and training should emphasise the hazard presented by poorly cleaned hands. Whilst the use of gloves can undoubtedly reduce the risk of cross-contamination from personnel themselves, it has been the experience of many companies that a policy of glove use encourages poorer hygiene practices because of the perceived barrier introduced between the individual and the food by the gloves. The correct use of gloves may be easier to control for employees dedicated to working at only one stage in a process such as placing meat slices onto sandwiches, but even in such situations, glove washing and changing/disposal procedures must be clearly defined to prevent abuse. Use of gloves in the retail environment, where a person may be serving customers with both ready-to-eat foods and raw meats, is likely to be less effective as the need for continual changing of gloves will soon lead some into the temptation of keeping the same gloves on for raw and cooked foods through the sheer inconvenience caused by the frequent changing procedures. In these cases, strict hand-cleansing procedures are more practical and effective.

Pests

Salmonella can be carried by a wide variety of pests including insects, such as cockroaches, rodents, such as mice and rats, and other animals, including reptiles and amphibians. The physical integrity of any manufacturing, retail and catering environment must be maintained to ensure that such pests do not gain entry to the building and cannot gain access to foods. This is achieved by effective biosecurity of the building to ensure that:

- walls, floors and drains do not have cracks or gaps
- doors are effectively protected
- windows are kept closed, unopenable or suitably screened
- ceilings are properly sealed, and
- ventilation systems are properly screened and filtered

Clearly, this must be done for both the inside structure of the building and the outside, as the entry of pests under floors or into wall cavities and roof

Table 4.47 Some questions to include in medical questionnaires for the purposes of enteric pathogen control

Question: Have you visited another country in the last month? If yes, please give details of the country, how long you were there and when you returned. *Consideration:* If a person has visited a country/continent where infectious enteric disease is more common, e.g. India, China, Africa, South America then microbiological screening of the individual prior to returning to work handling ready-to-eat foods may be prudent.
Question: Have you suffered from any vomiting or diarrhoea or been in contact with any person suffering from these symptoms in the last month? If yes, please give details of the symptoms, how long they lasted and when you (or the affected individual) stopped suffering from the symptoms. *Consideration:* Anyone suffering such symptoms should not handle ready-to-eat foods until they have been symptom free for 48 hours. It may also be prudent to exclude those who are in contact with people suffering these symptoms from handling ready-to-eat foods, i.e. only allow them to work in the low-risk side of a factory. Some companies choose also to request a stool sample for examination for *Salmonella* and clearance prior to return to work with ready-to-eat food.
Question: Have you been diagnosed as suffering from any gastrointestinal illness in the last month, e.g. salmonellosis, campylobacteriosis, shigellosis or dysentery? If yes, please give details of the illness and when it was diagnosed. *Consideration:* Anyone suffering suspected infectious intestinal illness of this nature should be excluded from work until they are free from symptoms for a period of 48 hours, e.g. after first normal stool (Anon, 1995a). Anyone suffering from confirmed infectious illness, e.g. salmonellosis, campylobacteriosis, *E. coli* gastroenteritis, should also be stool tested and not allowed to return to work handling ready-to-eat foods until a single normal stool sample is shown to be negative. When symptom free for 48 hours, people may return to work while awaiting results of stool testing but should not handle ready-to-eat or open food. For some illnesses, such as *E. coli* O157 infection and dysentery, an increased number of negative stool samples may be necessary.
Question: Have you ever been diagnosed as suffering from typhoid or paratyphoid? If yes, please give details of the illness and when it was diagnosed. *Consideration:* People suffering typhoid or paratyphoid should be excluded from work until at least six consecutive negative stool samples are obtained, each taken at 14 day intervals starting two weeks after the completion of antibiotic treatment (Anon, 1995a).

spaces will be a prelude to their appearance onto the factory floor. The use of insectocutors and suitably placed bait traps managed by professional third party pest control personnel are essential for the control of insect pests and vermin. The finding of increased levels of pests in traps or insectocutors must be quickly and actively responded to by the pest control contractors as the contamination of products by pests carrying *Salmonella* is a significant and real hazard to the majority of food processes. In addition, all waste containers and waste holding areas must be kept secure from invasion by insect pests and vermin.

Raw products where *Salmonella* may be present and customer cooking is designed to eliminate the hazard

A variety of products are purchased by retail customers that will, on occasion, be contaminated with enteric pathogens such as *Salmonella*. Raw chicken and raw meats and prepared, raw foods containing such ingredients are good examples. The ultimate safety of the product is then dependent on the customer handling and cooking the food effectively prior to consumption. Inadequate cooking and cross-contamination in the home, particularly from raw poultry, are well-known means by which people contract salmonellosis. It is therefore incumbent on the food industry, both manufacturers and retailers, to ensure that adequate guidance is given to consumers on appropriate handling and cooking methods for the product in question. As this is so important, it is essential that the cooking instructions are generated to reflect the actual cooking conditions likely to be available to the consumer.

There are a wide variety of types of domestic cooking appliances and together with the different cooker heat sources available, e.g. gas versus electric versus solid fuel, the cooker type and operation can have significant impact on the effectiveness of the cooking process. Cooking instructions should therefore be designed using cooking validation studies similar to the approach used by cooked product manufacturers. Consideration must be given, as applicable, to the size of the product, the number of products, the ingoing temperature (ambient, chilled or frozen), the cooking temperature, turning frequency, distance from the heat source, power setting of the heat source and the duration of the cook. As there are so many variables that can affect a cooking process, it is not surprising that the results of cooking can be so variable in the home. As much guidance as possible should be given on the product pack about effective cooking practices. Cooking advice should also make clear the variability that can occur in cooking processes. To help consumers judge whether or not effective cooking has been achieved, it is now common for manufacturers or retailers to place additional descriptive advice on the pack. However, the usefulness of some visual checks advised for use as an indication of effective cooking has been questioned by some studies that have demonstrated a poor relationship between internal cooking temperatures and visual changes in the internal colour of the product. A study by Lyon *et al.* (2000) found a large percentage of beefburgers had no pink colour (an indicator often used to judge cooking efficacy) even though the internal product temperature was not sufficient to destroy enteric contaminants, reinforcing the advice given in the USA to use temperature probes to judge cooking efficacy rather than visual means. It is still felt that advice given in the UK using phrases such as 'ensure the product is cooked until it is piping hot throughout and the juices run clear' can be useful simple guides for consumers to assess the effectiveness and safety of a cooking process.

Advice for susceptible groups

Some groups of products will always carry with them a greater risk of the presence of *Salmonella* than others no matter what controls are implemented and therefore these products will represent a greater potential risk to public health. In these cases, it is necessary to provide a suitable warning about the elevated risk to the consumer.

The potential presence of *Salmonella* and other enteric pathogens on raw meat and poultry products has prompted many processors in the UK to voluntarily label products with appropriate guidance on handling them to avoid cross-contamination. In the USA, it is a mandatory requirement to include food safety messages on the packaging of raw meat products, highlighting to consumers that they are handling a product potentially contaminated with pathogenic microorganisms.

Outbreaks of infection associated with unpasteurised fruit juice in the USA has prompted the US government to require processors of unpasteurised juice to label products 'WARNING: This product has not been pasteurised and, therefore, may contain harmful bacteria that can cause serious illness in children, the elderly, and persons with weakened immune systems' (Food and Drug Administration, 1998) if they cannot demonstrably achieve a defined reduction in enteric pathogens during the production process.

In a similar approach, outbreaks of salmonellosis implicating bean sprouts prompted the US government to advise consumers to avoid the consumption of raw sprouts and only to eat them after cooking (Food and Drug Administration, 1999b).

In the UK, concerns relating to the increased incidence of *Salmonella* in eggs and increases in infections caused by *S.* Enteritidis PT4, led the Department of Health to issue advice to the public to avoid the consumption of raw eggs or uncooked foods containing raw eggs (Anon, 1988a,b). This was extended to advice in relation to vulnerable people such as the elderly, the sick, babies and pregnant women to consume only eggs that are cooked until the white and yolk are solid. Such advice also appears voluntarily on the label on some boxes of retail eggs.

In the UK, some retailers of raw-milk, mould-ripened, soft cheeses also choose to label products with advice about the potential presence of harmful microorganisms in raw milk cheese and the particular need for vulnerable groups to avoid their consumption.

The objective of any advice of this nature must be to provide the consumer with sufficient information to make an informed choice about the purchase, handling or consumption of the product.

5

INDUSTRY ACTION AND REACTION

INTRODUCTION

Salmonella is probably the most common bacterial pathogen specified in microbiological criteria for foods appearing in a variety of regulatory criteria, industry guidelines and purchase specifications.

A microbiological criterion consists of statements concerning the micro-organism or microbial toxin of concern, the specific food and sample type to be examined, the sampling plan to be used, the test method to be used (the method must have been validated for the microorganism or toxin of concern in the food being examined), and the microbiological limit(s) to be applied indicating the interpretation to be placed on the result and a reaction procedure for those results which are in excess of the upper limit set. There are a variety of texts available which address these topics in detail (National Research Council, 1985; International Commission on Microbiological Specifications for Foods, 1986; Codex Alimentarius Commission, 1996a).

For the food industry, microbiological criteria fall into three categories:

(1) *Standards*. These are microbiological criteria contained in a law. Compliance is always mandatory. Examples include most criteria in European Union (EU) Directives and Statutory Instruments of England and Wales. Standards are monitored by enforcement agencies.
(2) *Guidelines*. These are criteria applied at any stage of the food pro-duction and distribution system to indicate the microbiological condition of a sample. They are for management information and can assist in the identification of potential problem areas when used in trend analysis.
(3) *Specifications*. These are microbiological criteria applied to

individual raw materials, ingredients or the end product. They are commonly used in purchase agreements between the vendor and purchaser of a raw food material or finished food product.

LEGISLATION AND STANDARDS

The general approach taken to legislation in Europe and North America in the context of food safety is to indicate the clear responsibility of food business proprietors to produce and supply safe and wholesome foods. For instance, in the UK, The Food Safety (General Food Hygiene) Regulations, 1995 (Anon, 1995d) (which implements parts of the European Union Directive 93/43/EEC of 14th June 1993 on the hygiene of foodstuffs), Section 4(1) (Anon, 1993b) states:

'A proprietor of a food business shall ensure that any of the following operations, namely, the preparation, processing, manufacturing, packaging, storing, transportation, distribution, handling and offering for sale or supply, of food are carried out in a hygienic way.'

Further, in Section 4(3):

'A proprietor of a food business shall identify any step in the activities of the food business which is critical to ensuring food safety and ensure that adequate safety procedures are identified, implemented, maintained and reviewed on the basis of the following principles –
(a) analysis of the potential food hazards in a food business operation;
(b) identification of the points in those operations where food hazards may occur;
(c) deciding which of the points identified are critical to ensuring food safety ("critical points");
(d) identification and implementation of effective control and monitoring procedures at those critical points; and
(e) review of the analysis of food hazards, the critical points and the control and monitoring procedures periodically, and whenever the food business's operations change.'

Clearly the potentially severe nature of illness caused by *Salmonella* makes this group of organisms an essential consideration in the hazard analysis of most food business operations. The use of a structured hazard analysis approach to identify means of controlling the organism in food production processes is likely also to generate many of the required controls for other foodborne pathogenic bacteria.

In addition to the general but important and necessary responsibility imposed by legislation on food business proprietors, other legislation (sometimes referred to as vertical legislation because it deals with a specific food in contrast to horizontal legislation which applies to generic food production controls, e.g. food hygiene), may also apply depending on the food type and business. A clear responsibility rests with all food business proprietors to know and understand which legislation applies to their business and ensure they are in compliance. Some legislation contains microbiological standards and compliance is therefore compulsory.

Salmonella is specifically included in standards in some food related legislation (Table 5.1), although in some cases, a generic statement concerning the microbiological safety requirements for a specific product type is included, e.g. all pathogens are required to be absent in mineral water (Anon, 1980) and in reference to 'all milk-based products', pathogenic microorganisms and their toxins 'must not be present in quantities such as to affect the health of consumers' (Anon, 1992b); these statements will, of course, include *Salmonella*. The European Directive on the quality of water (European Council, 1998) states, among other requirements, that 'Member States shall take the measures necessary to ensure that water intended for human consumption is wholesome and clean' and that this means, in part, that the water must be free from any microorganism and parasites and from any substances which, in numbers or concentrations, constitute a potential danger to human health. Again, *Salmonella* must be included in this 'catch-all' definition. Water is a major component in many food products and a clean water supply is also crucial for effective cleaning operations in food manufacturing processes for equipment, utensils and the general environment. The safety of the water supply with respect to bacterial pathogens including *Salmonella* is therefore paramount in any food production system.

Many raw foods including meat, poultry and eggs, milk, fish and shellfish and vegetables will be contaminated, on some occasion, with enteric pathogens. Whilst the aim of obtaining such raw material supplies with little or no contamination with *Salmonella* is desirable and the widespread application of criteria specifying the absence of *Salmonella* has focused producers' and processors' attention on the need to minimise their occurrence, their occasional presence must be expected. 'Zero tolerance' in raw foods is not a practical option and the penalty for the presence of *Salmonella* in raw foods should perhaps be positive encouragement to invest in appropriate systems to reduce incidence further rather than imposing some severe financial penalty, e.g. rejection of otherwise good food stocks. The ultimate goal must be to eliminate the hazard of

Table 5.1 Some examples of food related legislation in which *Salmonella* is included

European Union Directive	Product	Level of *Salmonella* specified in the standard*	Additional comment/action to be taken
Council Directive 94/65/EC laying down the requirements for the production and placing on the market of minced meat and meat preparations. (*Official Journal of the European Communities* 31.12.94 No. L368/10–31.)	Minced meat at the production site	Absent in 10 g, $n = 5$, $c = 0$	Unsatisfactory if any unit result shows presence in 10 g
	Meat preparations	Absent in 1 g, $n = 5$, $c = 0$	Unsatisfactory if any unit shows presence in 1 g
Council Directive 91/493/EEC laying down the health conditions for the production and the placing on the market of fishery products. (*Official Journal of the European Communities* 24.9.91 No. L268/15.) Also, Commission Decision 93/51/EEC on the microbiological criteria applicable to the production of cooked crustaceans and molluscan shellfish. (*Official Journal of the European Communities* 21.1.93 No. L13/11.)	Cooked crustaceans and molluscan shellfish at the production site	Absent in 25 g, $n = 5$, $c = 0$	Compulsory criteria; withhold from the market
Council Directive 91/492/EEC laying down the health conditions for the production and placing on the market of live bivalve molluscs. (*Official Journal of the European Communities* 24.9.91 No. L268/1–14.)	Live bivalve molluscs intended for immediate human consumption	Must not contain *Salmonella* in 25 g of mollusc flesh	Must comply

(Continued on p. 268)

Table 5.1 *Continued*

European Union Directive	Product	Level of *Salmonella* specified in the standard*	Additional comment/action to be taken
Council Directive 89/437/EEC on hygiene and health problems affecting the production and the placing on the market of egg products. (*Official Journal of the European Communities* 22.7.89 No. L212/87–100.) Sampled in treatment establishments.	All batches of egg products after treatment	Absent in 25 g or ml	Only products meeting the requirements may be used as foodstuffs or in the manufacture of foodstuffs
Council Directive 92/46/EEC laying down the health rules for the production and placing on the market of raw milk, heat-treated milk and milk-based products. (*Official Journal of the European Communities* 14.9.92 No. L268/1–34.) Criteria apply on removal from the processing establishment.	Raw cows' milk for drinking in that state	Absent in 25 g, $n = 5$, $c = 0$	When standard exceeded, the competent authority to investigate and take appropriate action
	Milk powder	Absent in 25 g, $n = 10$, $c = 0$	If standard exceeded, foodstuff must be excluded from human consumption and withdrawn from market
	All milk products except milk powder	Absent in 25 g, $n = 5$, $c = 0$	As for milk powder

* n = number of units making up the sample, c = number of units in the sample allowed to have a value outside the limit set in the standard.

Salmonella, but the realistic target for raw food and food ingredient producers and processors is the consistent minimisation of its occurrence.

This common sense view finds endorsement in a document prepared for the Codex Committee on Food Hygiene by the International Commission on Microbiological Specifications for Foods (Codex Alimentarius Commission, 1996b). In considering the control of *Salmonella* in foods, the Commission indicated that at present *Salmonella* cannot be eliminated from most farms, slaughtering processes, raw milk, fruits and vegetables, although good hygienic practices and properly implemented HACCP systems may minimise or reduce levels of contamination.

GUIDELINES

Microbiological guidelines used in industry are rarely published as they are generally developed in association with particular processes and products, consequently, a degree of confidentiality applies and industry guidelines remain in-house and self-imposed.

In the food manufacturing industry, results from routine tests for *Salmonella* indicating its presence, i.e. positive in 25 g samples (or other specified quantity), always lead to investigations to identify the source and rectify the cause.

General guidelines on the levels and types of microorganism relevant in specified foods produced under conditions of good manufacturing practice may be provided by industrial associations for their members, e.g. Institute of Food Science and Technology guidelines on the development and application of microbiological criteria for foods (Institute of Food Science and Technology, 1999). These include guidelines relating to *Salmonella* for a variety of product types and the maximum values ascribed to any ready-to-eat food category where the organism is included is invariably 'not detected' in a prescibed quantity of the food material. In the UK, the Public Health Laboratory Service (PHLS) has published microbiological guidelines (Gilbert *et al.*, 2000) to assist food examiners and enforcement officers in assessing the microbiological quality of foods and to indicate levels of certain types of bacterial contamination considered to be a potential health risk in ready-to-eat foods at the point of sale. In these guidelines, if *Salmonella* is detected in 25 g of any of the product groups described in the guidelines, then the food is considered 'unacceptable – potentially hazardous'. Although these guidelines have no statutory status, it is suggested that results from food samples falling in this latter category might form the basis for a prosecution by the Environmental Health Departments.

Useful sampling plans and recommended microbiological limits for *Salmonella* in relation to some foods in international trade have been published by the International Commission on Microbiological Specifications for Foods (1986). Some foods for which limits for *Salmonella* are considered for inclusion are cereal and cereal products, cocoa, chocolate and confectionery, dessicated coconut, dried meats and dried animal products, dried milk, eggs and egg products, frozen bakery products, frozen fruits, peanut butter, poultry and poultry products, processed meats, seafoods, soya flours, concentrates and isolates and yeast. These plans and sometimes the microbiological limits are often used by the food industry in normal business buying specifications. It is recognised however, that such microbiological testing has inherent limitations and that emphasis should be placed on preventative systems of bacterial pathogen control using HACCP-based approaches.

Some guidelines or codes of practice specifically targeting control of *Salmonella* published by government departments or industry trade bodies are shown in Table 5.2. In addition to these, there are numerous other guidelines and codes of practice dealing with good hygienic practice, HACCP and good manufacturing practice, or targeting specific food product types, e.g. ice cream, cheese and produce that contain some elements relating to control of enteric pathogens including *Salmonella* (Institute of Food Science and Technology, 1993). Indeed, some codes of practice were produced in specific response to public health issues arising concerning *Salmonella*, e.g. emulsified and non-emulsified sauces containing acetic acid (C.I.M.S.C.E.E., 1992), sprouted seeds (Brown and Oscroft, 1989; National Advisory Committee on Microbiological Criteria for Foods, 1999) and chocolate products (Collins-Thompson *et al.*, 1981) (Tables 5.2 and 5.3).

SPECIFICATIONS

Product specifications drawn up between a food manufacturer and customer (often a retailer), usually include information concerning the physical appearance of the product, physico-chemical characteristics of importance to the safety and/or quality of the product, e.g. pH, water activity, salt level, preservative level, and microbiological parameters also relevant to the safety and quality of the product.

Table 5.2 Some guidelines and codes of practice relating to control of *Salmonella* in materials related to food and food production, adapted from Institute of Food Science and Technology (1993), Advisory Committee on the Microbiological Safety of Food (1996) and Anon (2000i)

Title	Year	Source
Code of practice for the prevention and control of *Salmonella* in breeding flocks and hatcheries	1995	Ministry of Agriculture, Fisheries and Food (MAFF), London, UK
Code of practice for the control of salmonellae in commercial laying flocks	1988	
Code of practice for the control of *Salmonella* during the storage, handling and transport of raw materials intended for incorporation into, or direct use, as animal feedingstuffs	1989 Revised 1995	MAFF Publications PB2202
Code of practice for the control of *Salmonella* in the production of final feed for livestock in premises producing less than 10 000 tonnes per annum	1989 Revised 1995	MAFF, London, UK
Code of practice for the control of *Salmonella* in the production of final feed for livestock in premises producing over 10 000 tonnes per annum	1989 Revised 1995	
Code of practice for the control of *Salmonella* for the UK fishmeal industry	1989 Revised 1995	
Codes of practice for the control of *Salmonella* in the production of final feed for livestock	1995	MAFF Publications PB2200 and PB2201

(Continued on p. 272.)

Table 5.2 *Continued*

Title	Year	Source
Code of practice for the control of *Salmonella* in broilers	1989	MAFF, London, UK
Code of practice for the control of *Salmonella* in turkeys	1989	MAFF, London, UK
Code of practice for the prevention and control of *Salmonella* on pig farms	2000	MAFF, London, UK
Salmonella and related microorganisms in cocoa, chocolate and confectionery ingredients and products: Guidelines for good manufacturing practice	1984	Biscuit, Cake, Chocolate and Confectionery Alliance, London, UK
Code of hygienic practice based on HACCP for the prevention of *Salmonella* contamination in cocoa and chocolate and confectionery products	1991	International Office of Cocoa, Chocolate and Sugar Confectionery, Brussels, Belgium

Table 5.3 Some guidelines and codes of practice relating to practices that will help control bacterial pathogens including *Salmonella*

Publication	Year	Source
Guidelines for the hygienic manufacture, distribution and retail sale of sprouted seeds with particular reference to mung beans	1989	Brown and Oscroft (1989)
Code for the production of microbiologically safe and stable emulsified and non-emulsified sauces containing acetic acid	1992	Comité des Industries des Mayonnaises et Sauces Condimentaires de la Communauté Économique Européenne (C.I.M.S.C.E.E.)
The prevention of human transmission of gastrointestinal infections, infestations, and bacterial intoxications. A guide for public health physicians and environmental health officers in England and Wales	1995	Public Health Laboratory Service of England and Wales (Anon, 1995a)
Microbiological safety evaluations and recommendations on sprouted seeds	1999	National Advisory Committee on Microbiological Criteria for Foods

There has been a tendency in the recent past (still persisting in some areas of the industry) to establish and maintain a fixed list of micro-organisms (foodborne pathogens as well as non-pathogens, spoilage organisms and general microbiological tests such as total colony counts and coliforms), which is then included in all product specifications regardless of relevance to the product, the processes by which it was produced or the post-manufacturing conditions/treatments to which it will be exposed. Thus, in some food industry specifications, *Salmonella* is still being specified as a rejection criterion for raw meats and raw meat products intended for full cooking before consumption. It is generally inappropriate to specify *Salmonella* as a rejection criterion for raw meats but it is appropriate for ready-to-eat products (Gilbert *et al.*, 2000). It may be useful however, to monitor raw meats and poultry for the presence of the organism as part of trend analysis to ensure the incidence does not become excessive.

If, after consideration of all the issues relevant to determining where and how processes and products need to be monitored, microbiological testing and associated criteria are regarded as useful, then it is important to ensure that only those organisms relevant to the raw materials, processes and finished product are selected for testing purposes (Institute of Food Science and Technology, 1999). Where microbiological specifications are applied and it is considered necessary to include *Salmonella*, in addition to specifying the number of samples required to be taken, the target limit applied for example, to ready-to-eat foods, is often stated as 'not detectable in 25 g' with an 'unacceptable limit' of 'detection in 25 g' (Institute of Food Science and Technology, 1999).

For some product types destined for vulnerable groups such as very young children, product may be held by the manufacturer until finished product tests for *Salmonella* are completed and reported as satisfactory against the specification; this is often referred to as 'positive release' of finished product.

Confirmed detection of *Salmonella* in ready-to-eat products or those destined for vulnerable groups usually leads to complete removal of the products from the distribution and retail system, and on occasion, to public recall and closure of the manufacturing line or unit pending full investigation.

The size, number, composition and sampling method of food product samples will affect the outcome of a test and these should always be specified. Consistency with respect to these points is important to help maximise repeatability of results, but heterogeneous distribution of microorganisms can still cause results to vary significantly between tests, which is why on many occasions repeat sampling and testing of a batch of food product following a positive result (organism detected) produces a negative result (organism is not detected). The initial positive result must not be ignored. The number of positive versus negative results obtained can be used as an indication of the level of contamination present where this is relevant to decision making.

MONITORING FOR *SALMONELLA*

The success or otherwise of any system put in place to control pathogenic microorganisms including *Salmonella* is usually monitored by microbiologically examining samples taken from selected points, e.g. incoming raw materials, food materials in process such as after washing procedures, cooking procedures or slicing operations, and finished products.

It must be understood that microbiological examination of food products for organisms such as *Salmonella* which, if present, are often only in low numbers and distributed heterogeneously within a batch of food, is usually statistically invalid. In other words the chance of detecting the organism is likely to be very low. As such, reliance on microbiological testing for assuring food safety is never a sound approach. Alternative, non-microbiological methods of monitoring the efficacy of critical control points identified through HACCP are always advocated, including temperature and/or pH monitoring or similar simple, rapid, on-line monitors, preferably ones that can be carried out automatically and continuously. Microbiological testing should be seen as an additional means of assessing the efficacy of processing in keeping microorganisms under control but must only be applied with complete understanding of the inherent limitations of such testing.

The buying specification for some raw material supplies may include criteria to be met for *Salmonella,* e.g. 'not detected in 25 g' for prepared cooked meats for sandwich production, 'not detected in 250 g' for milk powder destined for baby food products, etc. Where a high degree of confidence is needed in the detection of very low numbers of *Salmonella* (if present) in a batch of product, a large number of samples must be taken (International Commission on Microbiological Specifications for Foods, 1986). For practical and economic reasons, it is common practice to combine a number of samples taken to reduce the number of tests required.

In cases where very large quantities, i.e. some tonnes of raw material, are involved, e.g. milk powder and egg powder, small sample quantities (a few grams) are taken at frequent intervals, using automatic in-line sampling devices. This significantly increases the chances of capturing the target microorganism, if present in the batch. These small quantities are bulked together and mixed before taking a test sample for examination in the laboratory. It is common for 'high risk' or 'sensitive' ingredients to be sampled and tested by the ingredient supplier prior to release to the purchaser. The certificate of analysis should then be supplied to the purchaser with the raw material batch. It is normal for the purchaser then to supplement this with some microbiological examinations of incoming batches of material. In this case, small portions (10–20 g) may be aseptically removed from several packages, combined and mixed prior to a relevant test portion being used in the laboratory examination.

Finished product samples are usually taken as complete finished packs from the end of the production line; single packs may also be taken at intervals during a production run. The numbers and frequency of sampling is usually in accordance with a customer's requirements. For monitoring purposes, it is often useful to take product samples from the start of the production run as any residual contamination on the equipment may be detected in the first batches of product from the line.

In the laboratory, it is important that a representative sample is removed from the pack for testing. For multicomponent products, this may involve selecting small quantities from each of the components in similar proportion to their percentage in the product so as to make up a final 25 g test portion. Alternatively, the test may be conducted on each individual component, or made up only from those components deemed to be the highest risk in respect of contamination. Again, the approach taken is usually in accordance with a customer's requirements but for most buying specification purposes, the first approach is used. The other options may be applied in problem solving situations, e.g. if *Salmonella* is found in the composite sample tested and a component analysis is warranted.

In some raw material and 'high-risk/high-care' food production units, e.g. milk powder, baby food, chocolate products, it is considered important to operate an environmental monitoring scheme that includes *Salmonella* as a target organism. For meaningful results to be obtained, samples taken must be relevant and representative of the area or material targeted. The objective of any sampling programme must be to find any contamination of specific concern so that the defined reaction measures can be focused to improve control in areas where the hazard is identified.

Environmental monitoring specifically targeting *Salmonella* generally requires samples that are large in volume or weight or from large areas (using sponge swabs). This is to ensure the sample is representative of the environmental area and to take account of the likely low level and incidence of the organism if it is present at all.

All sampling procedures, microbiological methods, criteria and reaction procedures to be followed in the event of the positive identification of *Salmonella* must be clearly indicated in standard operating procedures and the responsibilities of key personnel at all monitoring, reporting and control levels clearly defined. Table 5.4 indicates some approaches to action to be taken in response to results from tests carried out to detect *Salmonella* in environmental or ready-to-eat food samples. These would normally form part of an internal quality system.

Responsible food industry manufacturers and retailers regard *Salmonella* as a 'serious incident' in terms of their reaction to 'positive' results reports in ready-to-eat and 'sensitive' foods, e.g. baby foods. Consequently, well-defined procedures are in place in such companies for dealing with 'positives'. Some of the key lines of investigation, considerations and actions to be taken that may be included in these procedures are as follows:

Immediate action

(1) Organism related
- Establish that the result is confirmed as *Salmonella* and is not a 'presumptive' result.
- Establish full identification of the *Salmonella* serotype to assist with traceability.
- Establish whether the organism may be a laboratory cross-contaminant, e.g. is it the same as the positive control strain used in the laboratory or has the same strain been isolated from two completely different products sampled on the same day. The use of ribotyping techniques, available at some research facilities, to assess the relatedness of such strains can be useful in establishing this. Results from the application of this technique are produced in just a few hours.

(2) Product related
- Establish full details of the product(s) involved (production codes, location, consumer instructions, etc.) and determine any related products, e.g. by common ingredients, common production lines, etc.

Table 5.4 Considerations to be taken into account when *Salmonella* is detected in samples relating to ready-to-eat food production

Consideration	Not detected in any samples	Present in environmental samples	Present in test sample, e.g. 25 g raw materials	Present in test sample, e.g. 25 g ready-to-eat food
No action	Yes			
Environmental monitoring: increase sampling points to identify source		Yes	Yes	Yes
Cleaning efficacy: check procedures/cleaning records and monitor pre- and post-cleaning		Yes	Yes	Yes
Raw material testing: increase level of testing to identify frequency of contamination			Yes	Yes
Intermediate product testing: increase level of testing of in-process material to identify stages of contamination			Yes	Yes
Finished product testing: increase and include other products made in same area and with similar raw materials		Yes: all products in the area	Yes	Yes

Table 5.4 *Continued*

Consideration	Not detected in any samples	Present in environmental samples	Present in test sample, e.g. 25 g raw materials	Present in test sample, e.g. 25 g ready-to-eat food
Review efficacy of process controls: check all process records to ensure current controls have been carried out properly		Yes	Yes	Yes
Stop production: cease production until the problem is identified and resolved		*	†	Yes
Withdrawal of product: consider the need to withdraw products from sale or recall product from the public			†	Yes

Information given is for guidance only and may not be appropriate for individual circumstances.
*This may be considered necessary for some products such as baby food manufacture if the positive isolation is from the 'high-risk' side.
†This may be considered necessary if the material has been used in a ready-to-eat food where no reduction/destruction process is applied by the manufacturer.

- Instigate microbiological examination of additional product packs from the 'positive' batch to assess incidence of *Salmonella* in the batch and other production codes of the same product, ingredients, environmental samples, etc., to locate and assess any spread of the organism in the system.
- Establish microbiological histories of:
  - the 'positive' product,
  - raw materials - in-house checks and raw material supplier's checks,
  - related equipment and environmental hygiene records,
  - personnel - recent illness etc.
- Review customer complaint data to determine any evidence of increased complaints or complaints consistent with salmonellosis.
- Carry out a 'risk assessment' to determine the need for and extent of any product withdrawal from distribution or public recall.

Investigation and associated activities

- Identify equipment and staff involved with the implicated product.
- Consider,
  - other products made on the same production line (or in close proximity)/by the same people/with common ingredients,
  - equipment used: strip down and examine all relevant food contact equipment,
  - environmental sources: floors, drains, air ducts/filters, vacuum cleaners, mops/buckets/brushes used in cleaning, sieves, tray-wash, trays/trolleys/bins, wheels, boots/shoes, overhead pipes/girders/light fittings, etc., condensate, evaporators,
  - staff: hygiene checks, recent holidays and history of illness, stool checks, site facilities, e.g. canteen, toilets.
- Check all identified Critical Control Points have been monitored and correct records maintained, check records.
- Trace all raw materials used:
  - list all raw materials,
  - establish all sources, codes and delivery dates,
  - check certificates of analysis/supplier's microbiology reports *versus* buying specifications,
  - identify other products in which the same raw materials are used.
- Inspection and microbiological testing (where applicable) of:
  - process environment: floors, drains, air conditioning, refrigeration units,
  - process equipment: cleanability and effectiveness of cleaning, maintenance records,
  - process flow: control of areas of potential cross-contamination,

- hygiene schedules: properly maintained, correct chemicals and procedures used,
- personnel facilities: hand-wash units, toilets, changing rooms, canteen,
- biosecurity/pest control: birds, rodents, domestic animals.
- Considerations for factory clean-down (if applicable):
  - establish a logical flow for thoroughly cleaning the environment (floors, walls, ceiling) and manufacturing equipment with positive microbiological clearance of each area cleaned; include all over-heads, chiller and freezer evaporator units, drains, pipework and peripheral equipment such as ladders, engineering tools and cleaning equipment; check for any 'obscure'/'dead' areas,
  - cleaning of air-handling systems and changing of filters.

Production start up – precautions

- Positive clearance of high-risk raw materials with respect to *Salmonella*.
- Intensification of CCP checks.
- Increased sampling plan for monitoring the environment and finished product with respect to *Salmonella*.
- Increased monitoring of staff hygiene – hand swabs.

In any investigation by public health or environmental health enforcement officers of a case of foodborne salmonellosis, the ability of a food producer or retailer to demonstrate full traceability of events, processes, materials, etc. in relation to the product and well-structured and reliably operated procedures targeted to control *Salmonella* must be of considerable value. In addition, a well-structured plan for reacting to any untoward *Salmonella* 'incident' in the production system will enhance consumer confidence.

6

TEST METHODS

CONVENTIONAL METHODS

Salmonella is an important organism in the food industry and its presence detected in many types of food will lead to immediate action to find and eliminate the source. Production lines and whole production units, particularly of ready-to-eat foods, have been closed down because *Salmonella* was detected in a sample. Tests carried out on foods to isolate and identify the organism are invariably tests for detection in a specified quantity rather than for enumeration.

Methods for detecting the presence of *Salmonella* in food samples have been well established now for some decades. Currently, most National and International microbiological standard methods for the detection of *Salmonella*, and food industry recommended methods, are conventionally based, e.g. European Standard 12824 (Anon, 1998f); Association of Official Analytical Chemists (1995); UK Public Health Laboratory Service (Roberts *et al.*, 1995); Campden and Chorleywood Food Research Association (1997). Figure 6.1 illustrates the principles followed by most users of conventionally based methods for detecting *Salmonella* in foods, i.e. pre-enrichment of the sample, sub-culturing from the pre-enrichment broth into two selective enrichment broths which, after incubation, are each streaked onto two selective plating media. Any colonies suspected of being *Salmonella* observed on the selective agar plates are purified prior to confirmation, identification and further characterisation. When conducting any tests for confirming the identity of *Salmonella*, it is good practice always to use purified cultures.

There are a large number of variations to this basic method that have been introduced to account for the user's required method sensitivity and/or specificity or to deal with problems associated with the sample type. For instance, certain foodstuffs are inhibitory to *Salmonella* and alternative

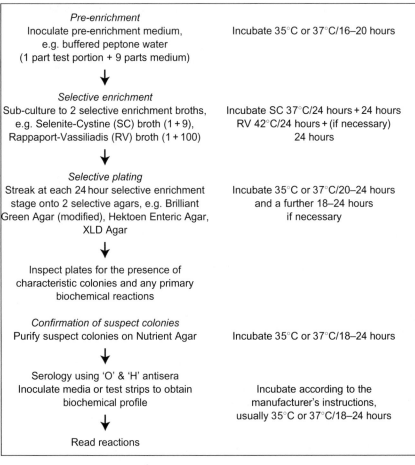

Figure 6.1 Conventional method for the isolation and identification of *Salmonella* spp.

pre-enrichment procedures or media may be used to improve test sensitivity; Table 6.1 indicates some food materials for which alternative pre-enrichment systems are used.

There is a considerable choice of media available for use as selective enrichment broths and selective plating media and yet more are in development. Table 6.2 lists just a few of these to indicate the variety that are commercially available. Some of the newer types of selective plating media employ chromogenic or fluorogenic substrates to enhance the specificity of the media, facilitate colony recognition and help reduce false positive information.

Table 6.1 Alternative pre-enrichment systems used for some food types

Food type	Pre-enrichment	Effect
Foods with a high fat content, e.g. cheese	Buffered peptone water with 0.22% Tergitol 7	Aids fat dispersion
Highly acidic or alkaline products	Adjust pH of pre-enrichment broth to 6.6–7.0 before incubation	Neutralises acid or alkali
Chocolate and confectionery products	Reconstituted skim milk powder (10% w/v) with Brilliant Green dye (final concentration of 0.002% w/v)	Reduces inhibition of *Salmonella*
Garlic and onion	Buffered peptone water with potassium sulphite (0.5% final concentration)	Reduces inhibition of *Salmonella*
Products that may contain inhibitory substances or products that may be osmotically active, e.g. some herbs and spices (oregano, cinnamon, cloves), honey	1:100 dilution of sample in buffered peptone water, e.g. 25 g + 2475 ml	Reduces inhibition of *Salmonella*

Table 6.2 Some more commonly used commercially available selective enrichment broths and selective plating media in conventional tests for the detection of *Salmonella*

Selective enrichment broths	Selective plating media
Rappaport-Vassiliadis Enrichment Broth	Brilliant Green Agar
Rappaport-Vassiliadis Soya Peptone Broth	MLCB Agar
Muller–Kauffmann Tetrathionate Broth	XLD Agar
Tetrathionate Broth	Bismuth Sulphite Agar
Selenite-Cystine Broth	Hektoen Enteric Agar
Selenite F Broth	Desoxycholate-Citrate Agar
	Salmonella Shigella Agar
	DLCS Agar
	Rambach Agar
	SM-ID Agar
	XLT4 Agar
	Harlequin *Salmonella* ABC
	Modified semi-solid Rappaport-Vassiliadis (MSRV) Agar

Table 6.3 indicates some of the modified conventionally based methods available and Table 6.4 lists some of the kits and systems available for characterising and identifying the suspect colonies isolated on the selective plating media and following purification on a non-selective medium. Although there are a variety of miniaturised and automated biochemical kits/systems for assisting in the identification of *Salmonella*, a number of common tests are used in these and Table 6.5 lists the key biochemical tests that may be applied to distinguish *Salmonella* from other members of the Enterobacteriaceae. It should be borne in mind that some strains occasionally exhibit atypical biochemical reactions. For example, although strains of *S. enterica* subsp. *diarizonae* (subgroup IIIb) more commonly ferment lactose, serotypes within *S. enterica* subsp. *enterica* (subgroup I) do not. However, occasionally, lactose fermenting strains of subgroup I are found such as the three serotypes, *S.* Anatum, *S.* Tennessee and *S.* Newington, that comprised the 86 (15.6%) lactose positive cultures of the total 552 *Salmonella* cultures isolated from dried milk products and milk-drying plants and examined in lactose fermentation tests (Blackburn and Ellis, 1973). This activity was also reported in strains of *S.* Senftenberg isolated from chicken giblets in 1974, *S.* Indiana associated with a particular turkey processing plant between 1977 and 1979 and *S.* Anatum isolated from an imported milk powder in 1980 (Threlfall *et al.*, 1983). It is prudent therefore, from time to time, particularly for unfamiliar sample types or different sources of milk-based raw materials, to employ combinations of selective plating media that exploit different biochemical characteristics of the organism, e.g. in the case of lactose fermenting *Salmonella*, include the use of a medium such as Mannitol Lysine Crystal Violet Brilliant Green (MLCB) Agar or Bismuth Sulphite Agar, that do not rely on lactose to differentiate *Salmonella* from other organisms growing on the same medium.

Because different strains of *Salmonella* are sensitive to different combinations of inhibitory substances and temperatures, the use of appropriately selected combinations of selective enrichment broths, selective plating media and incubation conditions provides the best confidence in results from tests for detecting *Salmonella*, if they are present in a sample.

As described, food microbiologists commonly use two selective enrichment broths and two selective plating media for routine testing purposes. However, for some foodstuffs and circumstances, e.g. new product development or trouble-shooting, additional procedures may need to be used. These may include the use of additional selective enrichment broths

Table 6.3 Some commercially available modified conventionally based methods for the detection of *Salmonella*

Test type	Name of test	Supplier
Motility, enrichment, immuno-immobilisation	1–2 test	Biocontrol Systems
Immunomagnetic separation	Dynabeads anti-*Salmonella*	Dynal (UK) Ltd
Semi-solid medium with selective agents	Diasalm	Merck (UK) Ltd
	Modified semi-solid Rappaport-Vassiliadis (MSRV) Medium	Oxoid Ltd
Modified growth system	Oxoid Rapid *Salmonella* test	Oxoid Ltd
Modified enrichment system	SPRINT enrichment	Oxoid Ltd
Membrane filtration and plate incubation	Iso-Grid	QA Life Sciences

Table 6.4 Some commercially available kits/methods for use in confirming/identifying *Salmonella*

Test type	Name of test	Supplier
Miniaturised biochemical test kits	API 20E	bioMérieux UK Ltd
	VITEK	bioMérieux UK Ltd
	Micro-ID®	Organon Teknika Ltd
	Microbact® Gram Negative Identification System 12A	Microgen Bioproducts Ltd
Latex agglutination	Microscreen *Salmonella*	Microgen Bioproducts Ltd
	Spectate	Rhône-diagnostics Technologies Ltd
	Pastorex	Sanofi Diagnostics Pasteur
Colorimetric enzyme	O.B.I.S. *Salmonella*	Oxoid Ltd
Molecular method – PCR	BAX™ for confirming *Salmonella*	Qualicon

Table 6.5 Some key biochemical tests commonly applied to identify suspect colonies as *Salmonella*

Characteristic	Usual reaction
Oxidase	−
Hydrogen sulphide production	+
Lysine decarboxylase	+
Ornithine decarboxylase*	+
Lactose fermentation	−
Sucrose fermentation	−
Glucose fermentation*	+
Mannitol fermentation*	+
Xylose fermentation*	+
Rhamnose fermentation*	+
Arabinose fermentation*	+
Hydrolysis of o-nitrophenyl-β-d-galactoside by action of β-galactosidase (ONPG)	−
Indole production	−
Urea hydrolysis	−
Acetoin production (Voges–Proskauer reaction)	−
Citrate used as sole carbon source*	+

+ = positive reaction; − = negative reaction.
*An important exception is *S*. Typhi which is negative in these tests.

and selective plating media and incubation of all broths and plates for 24 and 48 hours, sub-culturing from the broths to the plates after both 24 hours and 48 hours.

It is essential to ensure that all methods and variations used in the laboratory are validated for use with the products under examination and documented clearly for training and reference purposes.

ALTERNATIVE METHODS

As with methods for the detection of other foodborne bacteria, there is a need to supplement conventional methods with simpler, labour-saving methods preferably capable of delivering results more quickly and reliably than is possible using conventional methods. This is particularly important with respect to *Salmonella* and ready-to-eat foods and foods for vulnerable groups, e.g. very young children, because of the potentially serious public health and commercial consequences of a positive result in such foods. Rapid screening methods facilitating positive release programmes for many food types are important in the food industry.

When selecting any method for use, all of the following attributes need consideration:

- sensitivity
- specificity
- simplicity
- robustness
- reliability
- 'hands-on' time
- the need for additional tests to confirm presumptive results, and time taken to obtain the final result
- requirement for trained staff and/or special equipment
- cost per test.

A variety of alternative techniques exist for the detection and/or identification of *Salmonella* including those based on the use of antibodies to *Salmonella* antigens, e.g. enzyme-linked immuno-sorbent assays (ELISA), immuno-chromatography, chemiluminescent immunoassays, antibody coated dipsticks or beads, latex agglutination and other technologies such as electrical conductance methods and the polymerase chain reaction (PCR). The list continues to grow as new technologies are exploited for application in microbiology. Regular updates on commercially available alternative microbiological test methods are available together with useful key references and the validation status of the method (Baylis, 2000).

Many alternative tests depend on the presence of a minimum number of target cells for reliable detection of the target organism, i.e. *Salmonella*, and the procedures described for use with a specific manufacturer's kit or test are designed to assure a reliable result. To avoid false negative or false positive results, it is always important to read and understand the technical information supplied by the kit/test manufacturer and carefully follow the instructions for carrying out the test. In selecting methods for use, a clear understanding of the test's capability is necessary, i.e. whether it is a screening test or a test giving a confirmed positive result. In the former case, a positive result may require further work to be carried out to confirm and identify the organism responsible for the result. The procedures to be followed and action to be taken when a positive result is obtained from the test employed should be clearly defined in the laboratory's operating procedures.

Table 6.6 indicates some of the currently available alternative test methods for the detection or identification of *Salmonella* and the approximate time taken to obtain a result where this is applicable.

Biochemical identification kits

These are some of the simplest forms of labour-saving test systems for identifying *Salmonella*. They consist of a range of biochemical reactions (Table 6.5) produced in pre-formed chambers and supplied in a disposable unit. Following inoculation and incubation of the test chambers, reaction results are assessed, usually by noting and scoring specific colour changes in the media. Following assessment, a profile of the organism is obtained which is used for its identification. For *Salmonella*, it is necessary to supplement these biochemical test results with serological tests.

Enzyme-linked immuno-sorbent assay (ELISA)

The success of ELISA methods is based on the high specificity of the antibody used to its target antigen, i.e. *Salmonella*. Jones (2000) gives an excellent description of the principles underlying and the types of anti-body-based methods used. The anti-*Salmonella* antibodies are bound to a solid substrate, e.g. the internal surface of the wells in a microtitre plate, and these are used to capture *Salmonella* antigens present in a treated enrichment broth placed in the well. Following a sequence of manipulations involving washings, addition of further reagents and incubations, a coloured end product is obtained in those wells containing *Salmonella* antigens. Initial sample enrichment based on the use of conventional selective enrichment broths is necessary to ensure sufficient target cells are available (10^4-10^6/ml depending on test type used) for detection in a contaminated sample.

Positive results obtained using ELISA methods are usually confirmed by streaking the original enrichment broth onto selective agar and then following the conventional approach to confirm the presence and specific type of *Salmonella* found.

ELISA methods can offer rapid screening of samples with a negative result available in 24 hours and an early indication of a potential positive result.

Table 6.6 Examples of alternative methods available for screening for the presence of *Salmonella*, adapted from Baylis (2000)

Test type	Name of test	Approximate test time (hours)	Supplier
Electrical conductance	Bactometer	42 maximum	bioMérieux UK Ltd
	Malthus	42 maximum	Malthus Instruments
	RABIT	42	Don Whitley Scientific Ltd
Enzyme-linked immuno-sorbent assay (ELISA)	Vidas *Salmonella*	26	bioMérieux UK Ltd
	Salmonella-Tek	52	Organon Teknika Ltd
	TECRA *Salmonella* Visual Assay	42–52	TECRA Diagnostics, UK
	Transia Plate *Salmonella*	52	DiffChamb, S.A.
	EIAFOSS	26	Foss UK Ltd
	Locate	52	Rhône-diagnostics Technologies Ltd
Chemiluminescent immunoassay	ISO Screen *Salmonella*	24	Stratecon Diagnostics
Immuno-chromatography	Path-Stik	56	Celsis International plc
	REVEAL® for *Salmonella*	21	Neogen Corporation
Immunoprecipitate	Visual Immunoprecipitate Assay (VIP) for *Salmonella*	24/48	Bio Control Systems Inc.
Ice-nucleation	BIND® (Bacterial Ice Nucleation Detection) *Salmonella*	22	Bio Control Systems Inc.
Nucleic Acid Hybridisation Probe	Gene Trak *Salmonella* Assay	48	Gene Trak Systems
Polymerase chain reaction (PCR)	Foodproof® *Salmonella*	24	BioteCon Diagnostics
	TaqMan® for *Salmonella*	24	PE Applied Biosystems
	Probelia™	24	Sanofi Diagnostics Pasteur
	BAX™ for screening *Salmonella*	24	Qualicon

Immunomagnetic separation

This technique is used to improve the sensitivity and specificity of conventional selective methods for detecting the presence of *Salmonella* in samples. Small ($< 100\,\mu m$) magnetic beads, coated with antibodies to *Salmonella*, are used to specifically capture cells of the organism from enrichment broths. The magnetic property of the beads is used to separate the cells from the broth and these are plated out onto selective agar and the conventional approach then followed to isolate and identify the organism present.

Latex agglutination tests

Antibody coated latex beads are used in agglutination tests to provide a rapid indication of the presence of colonies of *Salmonella* on selective agar culture plates. Such tests offer specificity and time saving which can be important to a food microbiologist examining products against specifications.

Immuno-chromatography

In these tests, anti-*Salmonella* 'capture' antibody is immobilised in a line on a membrane located in a 'reaction' zone on a chromatographic pad unit or membrane. Generally, cultured enrichment broth is heat treated before being added to *Salmonella* specific 'detection' antibody bound to labelled particles (colloidal gold, coloured latex, etc.) dispersed in the first 'reaction' stage towards the bottom of the membrane. If *Salmonella* cells are present, the particles bind to the organisms via the detection antibody and they are drawn up the membrane. The immobilised antibody captures the *Salmonella*-antibody-particle complex and, as the particles are concentrated together, a coloured line forms indicating a positive result. Like all antibody-based tests, immuno-chromatographic tests may offer specificity and time saving which can be important in positive release product testing programmes, but positive results do require confirmation using cultural techniques.

Electrical techniques

Electrical techniques for the detection of microorganisms are based on the ability of instruments to monitor and detect small changes in the electrical properties of a medium in which a microorganism is growing. Specific broth media have been developed which favour the growth of *Salmonella*. As the organism grows, particular substrates in the broth are utilised, yielding products of metabolism that cause changes in the medium giving detectable changes in conductance. The changes that occur are recorded as a conductance curve. The curve generated is monitored by computer; if it meets certain pre-set criteria, the sample is registered as positive and the broth is considered to be presumptive positive for *Salmonella*. Presumptive positive broths are streaked out onto selective agar and the conventional method is followed to isolate and identify the organism present.

Ice-nucleation technique

This method employs bacteriophage that have been modified to carry DNA for the ice-nucleation protein. Tubes are supplied that contain bacteriophage and an indicator dye. Overnight pre-enrichment culture is added to the tube and the tube is incubated for 2–5 hours. If *Salmonella* is present, the bacteriophage attach to the cells and insert the modified DNA which causes the *Salmonella* to produce the ice-nucleation proteins. These promote the formation of ice crystals when the sample is cooled to a specific temperature. Positive samples freeze and turn orange but also require confirmation using cultural techniques applied to a separate, retained enrichment broth.

Polymerase chain reaction (PCR)

The polymerase chain reaction is a technique used to amplify the number of copies of a pre-selected region of DNA to a sufficient level to test for identification. For PCR-based tests for *Salmonella* in foods, the region of DNA selected is sequence specific to *Salmonella*. Jones (2000) gives a clear description of the principles of such molecular methods used in the microbiological examination of food. There are commercially available PCR-based screening methods for the detection of *Salmonella* in foods, however, although they provide results more rapidly than conventional methods so will be useful for quickly screening out negative samples and detecting contaminated samples, like some other alternative (to conventional) methods, any positive results need further work to characterise the strain found.

'Fingerprinting' methods

Once *Salmonella* has been isolated and identified using conventional or alternative methods, this level of identity together with the associated sample information is usually sufficient for a food microbiologist to interpret results against the original requirements for the test, e.g. a product buying specification. However, because of the importance of *Salmonella* in the food industry, it may be necessary to further discriminate the identity of the strain isolated, particularly if the origin of the strain needs to be traced in connection with cases of suspected illness.

The ability to subtype or 'fingerprint' *Salmonella* serotypes is important in their surveillance as well as in the investigation of outbreaks of illness and traceability in food processing and environmental/ecological situations. Techniques are available which can allow confident traceability of strains in the factory environment. There are a number of methods used for sub-classifying *Salmonella* include biotyping, serotyping (including variation in H antigens), phage typing, antibiotic resistance patterns (resistotyping), various molecular typing methods including pulsed-field gel electrophoresis (PFGE) (Old and Threlfall, 1998) and ribotyping, which is basically restriction fragment length polymorphism (RFLP) analysis of a specific set of microbial genes carrying the information for ribosomal RNA (Jones, 2000). This latter technique is automated and is quite commonly used to generate comparative DNA profiles of microbial strains in problem-solving investigations and strain tracing studies. When this level of differentiation is required, it is recommended that the services of national public health sector or research institute/association experts in these techniques and the interpretation of the results are obtained.

It is always important to ensure that the method used will reliably detect any *Salmonella* that may be present in the food being examined. It may be necessary to initially confirm by experiment that a particular method will be suitable for the purpose intended. Subsequently, in routine use, laboratory-quality assurance systems (internal and external) should be employed to verify ongoing test efficacy.

A variety of sources exist which may be used for guidance in appropriate method selection. National and International Standard methods published, for example, by the British Standards Institution, the European Committee for Standardisation (CEN), International Organisation for Standardisation (ISO), and the International Dairy Federation (IDF) are available. In addition, methods have been reviewed, practised and validated by reputable bodies such as the Public Health Laboratory Service (UK), the Association of Official Analytical Chemists (USA) and Campden & Chorleywood Food Research Association (UK).

7

THE FUTURE

Various studies have been carried out to assess the cost to society of outbreaks of foodborne illness including salmonellosis. Costs are incurred by individuals, employers and health services and these are mostly avoidable. Significant benefits can be gained in all of these areas from simple interventions such as pasteurisation of milk (Cohen *et al.*, 1983).

Structured hazard analysis of food production processes from primary agriculture and aquaculture throughout the chain to the consumer, if carried out properly, can identify the critical points that, if consistently well controlled, will prevent a large proportion of the cases of salmonellosis that occur.

In addition to the current and well-founded concern about levels of foodborne salmonellosis, the suggestion that genes facilitating virulence were introduced into *Salmonella* in gene transfer events from related bacterial genera, e.g. *Escherichia coli*, or between strains of *Salmonella* and that further gene transfer/acquisition could result in the emergence of new pathogens (Groisman and Ochman, 1997 and 2000) should underline the urgent need for more concerted action to prevent the preventable. This action should be led by governments.

It is important that governments continue to develop working partnerships with industry globally and at all levels to co-ordinate the production and dissemination of consistent and correct food safety information in relation to production, processing and practices. Such information needs to be readily available to food producers, processors, caterers, retailers, all food handlers and the public.

Simple but effective interventions, such as proper personal hygienic practices, prevention of extraneous contamination of foods, use of appropriate storage conditions, effective application of pasteurisation processes, etc., should be encouraged and where necessary, enforced by legislation. Success in these areas must contribute to a tangible reduction in cases of salmonellosis caused by the old and any new *Salmonella* serotypes that may 'emerge'.

For *Salmonella*, the implications of commercial developments in the food production industry are similar to those described for Vero cytotoxigenic *E. coli* (Bell and Kyriakides, 1998) and research relating to the sources, growth and survival characteristics of *Salmonella*, detection methods and effective, practical control measures that can be applied to these organisms in the food industry is still useful and must form a part of both short and mid-term future work programmes.

As described in this book, the ubiquitous nature of *Salmonella* and the wide diversity of foods that have been associated with outbreaks of salmonellosis should be enough of a warning to food scientists that *Salmonella* is a potential hazard to any food production process. Due consideration must be given to the organism in any hazard analysis of a food production process. Even though a great deal more work will undoubtedly be done to further build our knowledge and understanding of the routes of transmission and behaviour of different *Salmonella* serotypes in different conditions found in foods, the consistent and correct application of simple interventions will undoubtedly help reduce the hazard of *Salmonella* in all foods.

It is therefore, and will remain, essential that a detailed and competent hazard analysis is carried out during the primary stages of all new food product and process developments to ensure that relevant critical controls and monitoring systems are put in place and operated effectively. This will help to minimise potential public health problems which could arise from the presence and outgrowth of all types of *Salmonella*, whether 'old' or 'new'.

GLOSSARY OF TERMS

Biotyping The conventional method for distinguishing between bacterial types using their metabolic and/or physiological properties.

Commensal A relationship between two organisms from which one derives benefit and the other (the host) derives no benefit but is not harmed.

Confirmed positive A level of identification obtained following the use of specific diagnostic tests which allow full characterisation and clear identification of the organism according to accepted definitions.

D value The time required (usually expressed in minutes) at a given temperature to reduce the number of viable cells or spores of a given microorganism to 10% of the initial population.

Facultative Optional lifestyle; associated with mode not normally adopted, e.g. a facultative anaerobe usually grows aerobically but can grow anaerobically.

HACCP A system which identifies, evaluates and controls hazards which are significant for food safety (as defined by the Codex Alimentarius Commission in ALINORM 97/13A, CAC, Rome). Conventionally understood as 'Hazard Analysis Critical Control Point' but which may be more meaningfully interpreted as 'Hazard Analysis allowing Control of Critical Points in a food process'; it is a structured approach to identifying microbial hazards in a food process and defining and implementing systems for their control.

Hazard A biological, chemical or physical agent in, or condition of, food with the potential to cause an adverse effect on health (as defined by the Codex Alimentarius Commission in ALINORM 97/13A, CAC, Rome).

Infection In relation to bacterial food poisoning is a condition in which the pathogen multiplies in the host's body and becomes established in or on the cells or tissue of the host.

Pasteurisation   A form of heat treatment that kills vegetative pathogens and spoilage microorganisms in milk and other foods, e.g. for milk a common pasteurisation process is at least 71.7°C for 15 seconds.

Pathogen   A biological agent which, via direct infection of another organism, e.g. man, animals, etc., or toxin production causes disease in that organism.

Phage typing   A method used to distinguish between bacteria within the same species on the basis of their susceptibility to a range of bacterial viruses (bacteriophage).

Phenotype   The observable characteristics of an organism which include biotype, serotype, phage type and bacteriocin type.

Presumptive positive   A level of identification obtained following the use of primary diagnostic tests which allow characterisation of the organism according to accepted definitions; presumptive positives require further tests to be applied to confirm that the identity of the organism is as suspected.

Risk   An estimate of the likelihood of the hazard occurring and its potential adverse health effects.

Saprophyte   A plant-like bacterium or fungus which obtains its nutrients from dead and decaying organic (plant or animal) material.

Sequelae   A secondary illness or ailment occurring as the result of a previous disease/infection, e.g. a sequelae to salmonellosis may be reactive arthritis.

Serotyping   A method of distinguishing bacteria on the basis of their antigenic properties (reaction to known antisera). The O-antigen defines the *serogroup* of a strain and the H-antigen defines the *serotype* (serovar) of the strain, therefore, a number of serotypes may constitute a serogroup.

Strain   An isolate or group of isolates that can be distinguished from other isolates of the same genus and species by either phenotypic and/or genotypic characteristics.

Taxonomy   The biological classification/grouping of organisms according to their similarities.

Water activity (a_w)   A measure of the availability of water for the growth and metabolism of microorganisms. It is expressed as the ratio of the water vapour pressure of a food or solution to that of pure water at the same temperature.

z value   The increase in temperature (°C) required to decrease the D value by ten-fold.

REFERENCES

Adams, M.R., Hartley, A.D. and Cox, L.J. (1989) Factors affecting the efficacy of washing procedures used in the production of prepared salads. *Food Microbiology*, **6**, 69–77.

Advisory Committee on the Microbiological Safety of Food (ACMSF) (1992) *Report on vacuum packaging and associated processes*. HMSO, London, UK.

Advisory Committee on the Microbiological Safety of Food (ACMSF) (1993) *Report on* Salmonella *in eggs*. HMSO, London, UK.

Advisory Committee on the Microbiological Safety of Food (ACMSF) (1996) *Report on poultry meat*. HMSO, London, UK.

Ahmed, S. (1997) An outbreak of *E. coli* O157 in Central Scotland. *Scottish Centre for Infection and Environmental Health Weekly Report*, Number 1, No. 97/13, 8. Scottish Centre for Infection and Environmental Health, Glasgow, Scotland.

Aldsworth, T.G., Sharman, R.L., Dodd, C.E.R. *et al.* (1998) A competitive microflora increases the resistance of *Salmonella typhimurium* to inimical processes: evidence for a suicide response. *Applied and Environmental Microbiology*, **64**(4), 1323–1327.

Altekruse, S.F., Timbo, B.B., Mowbray, J.C. *et al.* (1998) Cheese-associated outbreaks of human illness in the United States, 1973–1992: Sanitary manufacturing practices protect consumers. *Journal of Food Protection,* **61**(10), 1405–1407.

Anderson, E.S., Ward, L.R., de Saxe, M.J. *et al.* (1977) Bacteriophage-typing designations of *Salmonella typhimurium*. *Journal of Hygiene*, Cambridge, **78**, 297–300.

Andrews, W.H., Mislivec, P.B., Wilson, C.R. *et al.* (1982) Microbial hazards associated with bean sprouting. *Journal of the Association of Official Analytical Chemists*, **65**(2), 241–248.

Anon (1975) *Salmonella typhimurium* outbreak traced to a commercial apple cider – New Jersey. *Morbidity and Mortality Weekly Report*, **24**, 87–88.

Anon (1980) Council Directive 80/777/EEC on the approximation of the laws of the member states relating to the exploitation and marketing of natural mineral waters. *Official Journal of the European Communities*, 30.8.80, No. L229/1.

Anon (1986) Foodborne disease surveillance in England and Wales 1984. *Communicable Disease Report*, **34**, 3–6.

Anon (1988a) Salmonella *and eggs*. Department of Health, London, UK. 21 November 1988.

Anon (1988b) *Chief medical officer repeats advice on raw egg consumption.* Department of Health, London, UK. 5 December 1988.

Anon (1992a) *Safer cooked meat production guidelines. A 10 point plan.* Department of Health, London, UK.

Anon (1992b) Council Directive 92/46/EEC laying down the health rules for the production and placing on the market of raw milk, heat-treated milk and milk-based products. *Official Journal of the European Communities*, 14.9.92, No. L268/1–34.

Anon (1993a) Surveillance Group Update: *Salmonella* contamination of foods. *PHLS Microbiology Digest*, **10**(2), 105.

Anon (1993b) Council Directive 93/43/EEC of 14th June 1993 on the hygiene of foodstuffs. *Official Journal of the European Communities*, 19.7.93, No. L175/1–11.

Anon (1994) *Nationwide beef microbiological baseline data collection program: Steers and heifers. October 1992–1993.* United States Department of Agriculture, Food Safety and Inspection Service, Washington, USA.

Anon (1995a) The prevention of human transmission of gastrointestinal infections, infestations, and bacterial intoxications. A guide for public health physicians and environmental health officers in England and Wales. *Communicable Disease Report Review*, **5**(11), R158–R172.

Anon (1995b) *A national study on ready to eat meats and meat products, Part 1*. Ministry of Agriculture, Fisheries and Food, London, UK.

Anon (1995c) *Biological control, rodents.* Promed@usa.healthnet.org (5 Dec 95).

Anon (1995d) *The Food Safety (General Food Hygiene) Regulations, 1995.* Statutory Instrument No. 1763. HMSO, London, UK.

Anon (1996a) *Salmonella gold-coast. Communicable Disease Report Weekly*, **6**(51), 443.

Anon (1996b) Salmonella gold-coast *and Farm Cheese. Food Hazard Warning.* Department of Health, London, UK. 18 December 1996.

Anon (1996c) *A national study on ready-to-eat meats and meat products. Part 3*. Ministry of Agriculture, Fisheries and Food, London, UK.

Anon (1997a) Preliminary report of an international outbreak of *Salmonella anatum* infection linked to infant formula milk. *EuroSurveillance Monthly*, **2**(3), 1–4.

Anon (1997b) *Salmonella gold-coast* and cheddar cheese: update. *Communicable Disease Report Weekly*, **7**(11), 93 and 96.

Anon (1997c) *Safer Eating: Microbiological food poisoning and its prevention.* Parliamentary Office of Science and Technology, London, UK.

Anon (1997d) *A national study on ready-to-eat meats and meat products. Part 4*. Ministry of Agriculture, Fisheries and Food, London, UK.

Anon (1997e) *Forward programme for the poultry industry. Report of the Government/Industry working group on meat hygiene.* Ministry of Agriculture, Fisheries and Food. September 1997.

Anon (1997f) Get Cracking. *Sainsbury's The Magazine*. November 1997.

Anon (1998a) *A Review of Antimicrobial Resistance in the Food Chain*. A Technical Report for MAFF. Ministry of Agriculture, Fisheries and Food, London, UK.

Anon (1998b) *Surveillance of the microbiological status of raw cows' milk on retail sale.* Microbiological Food Safety Surveillance, Department of Health, London, UK.

Anon (1998c) Defective pasteurisation linked to outbreak of *Salmonella typhimurium* definitive phage type 104 infection in Lancashire. *Communicable Disease Report Weekly*, **8**(38), 335 and 338.

Anon (1998d) Multi-state outbreak of *Salmonella* serotype Agona infections linked to toasted oats cereal – United States, April–May, 1998. *Morbidity and Mortality Weekly Report*, **47**(22), 462–464.

Anon (1998e) *Expert advice repeated on* Salmonella *and raw eggs*. Department of Health, London, UK. 9 April 1998.

Anon (1998f) European Standard EN 12824:1997. *Microbiology of food and animal feeding stuffs – Horizontal method for the detection of* Salmonella. British Standards Institution, London, UK.

Anon (1999a) *Gastrointestinal Infections*. Annual Review of Communicable Diseases – England and Wales 1997. Public Health Laboratory Service, Communicable Disease Surveillance Centre, London, UK.

Anon (1999b) *Salmonella java* phage type Dundee – rise in cases. *Communicable Disease Report Weekly*, **9**(9), 77.

Anon (1999c) *Salmonella java* phage type Dundee – rise in cases: update. *Communicable Disease Report Weekly*, **9**(12), 105 and 108.

Anon (1999d) Salmonellosis outbreak, South Australia. *Communicable Diseases – Australia*, **23**(3).

Anon (2000a) *PHLS* Salmonella *Fact Sheet* (updated 8 February 2001). www.phls.co.uk

Anon (2000b) *A report of the study of infectious intestinal disease in England*. The Stationery Office, London, UK.

Anon (2000c) National increase in *Salmonella typhimurium* DT104 – update. *Communicable Disease Report Weekly,* **10**(36), 323 and 326.

Anon (2000d) Outbreak of salmonellosis associated with chicks and ducklings at a children's nursery. *Communicable Disease Report Weekly*, **10**(17), 149 and 152.

Anon (2000e) *State Health Director advises consumers to wash cantaloupe before eating*. Press Release no. 29-00, 23 May 2000. California Department of Health Services, California, USA.

Anon (2000f) *Salmonellosis outbreak associated with raw mung bean sprouts*. Press Release no. 22-00, 19 April 2000. California Department of Health Services, California, USA. www.dha.ca.gov

Anon (2000g) *Report on the national study of ready-to-eat meats and meat products*: Part 5. Food Standards Agency, London, UK.

Anon (2000h) *The British Egg Information Service*. www.britegg.co.uk

Anon (2000i) *Code of practice for the prevention and control of* Salmonella *on pig farms*. Ministry of Agriculture, Fisheries and Food (MAFF), London, UK.

Anon (2001) International surveillance network for the enteric infections – *Salmonella* and VTEC O157. www.phls.co.uk/international/Enter-Net/enter-net.htm

Association of Official Analytical Chemists (1995) *Bacteriological Analytical Manual*, 8th edn. Food and Drug Administration, Association of Official Analytical Chemists International, Arlington, Virginia, USA.

Barnes, G.H. and Edwards, A.T. (1992) An investigation into an outbreak of *Salmonella enteritidis* phage type 4 infection and the consumption of custard slices and trifles. *Epidemiology and Infection*, **109**, 397–403.

Barrile, J.C., Cone, J.F. and Keeney, P.G. (1970) A study of salmonellae survival in milk chocolate. *Manufacturing Confectioner*, **50**(9), 34–39.

Barrile, J.C., Ostovar, K. and Keeney, P.G. (1971) Microflora of cocoa beans before and after roasting at 150°C. *Journal of Milk Food Technology*, **34**(7), 369-371.

Baylis, C. (2000) *The Catalogue of Rapid Microbiological Methods, Review No. 1*, 4th edn. Campden and Chorleywood Food Research Association, Chipping Campden, UK.

Bell, C. and Kyriakides, A. (1998) E. coli: *A practical approach to the organism and its control in foods*. Blackwell Science Ltd, Oxford, UK.

Besser, T.E., Goldoft, M., Pritchett, L.C. *et al.* (2000) Multiresistant *Salmonella* Typhimurium DT 104 infections of humans and domestic animals in the Pacific Northwest of the United States. *Epidemiology and Infection*, **124**, 193-200.

Besser, R.E., Lett, S.M., Weber, J.T. *et al.* (1993) An outbreak of diarrhea and hemolytic uremic syndrome from *Escherichia coli* O157:H7 in fresh-pressed apple cider. *Journal of the American Medical Association*, **269**(17), 2217-2220.

Beuchat, L.R. (1992) Surface disinfection of raw produce. *Journal of Food Protection*, **12**(1), 6-9.

Beuchat, L.R. (1996) Pathogenic micro-organisms associated with fresh produce. *Journal of Food Protection*, **59**(2), 204-216.

Beuchat, L.R. (1997) Comparison of chemical treatments to kill *Salmonella* on alfalfa seeds destined for sprout production. *International Journal of Food Microbiology*, **34**, 329-333.

Beuchat, L.R., Nail, B.V., Adler, B.B. *et al.* (1998) Efficacy of spray application of chlorinated water in killing pathogenic bacteria on raw apples, tomatoes, and lettuce. *Journal of Food Protection*, **61**(10), 1305-1311.

Blackburn, B.O. and Ellis, E.M. (1973) Lactose-fermenting *Salmonella* from dried milk and milk-drying plants. *Applied Microbiology*, **26**(5), 672-674.

Boase, J., Lipsky, S., Simani, P. *et al.* (1999) Outbreak of *Salmonella* serotype Muenchen infections associated with unpasteurised orange juice – United States and Canada, June 1999. *Morbidity and Mortality Weekly Report*, **48**(27), 582-585.

Bolder, N.M. (1997) Decontamination of meat and poultry carcasses. *Trends in Food Science and Technology*, **8**, 221-227.

Bouvet, E., Jestin, C. and Ancelle, R. (1986) *Importance of exported cases of salmonellosis in the revelation of an epidemic*. Proceedings of the 2nd World Congress on Foodborne Infections and Intoxications, West Berlin, p. 303.

Brenner, D.J. (1984) Family 1. Enterobacteriaceae RAHN 1937, Nom. Fam. Cons. Opin. 15, Jud. Comm. 1958, 73; Ewing, Farmer, and Brenner 1980, 674; Judicial commission 1981, 104. In: *Bergey's Manual of Systematic Bacteriology, Volume 1* (eds N.R. Krieg and J.G. Holt), Williams & Wilkins, Baltimore, pp. 411 and 415.

Brown, K.L. and Oscroft, C.A. (eds) (1989) *Guidelines for the hygienic manufacture and retail sale of sprouted seeds with particular reference to mung beans*. Technical Manual No. 25, Campden Food & Drink Research Association, Chipping Campden, UK.

Bryan, F.L. and Doyle, M.F. (1995) Health risks and consequences of *Salmonella* and *Campylobacter jejuni* in raw poultry. *Journal of Food Protection*, **58**(3), 326-344.

Buchanan, R.L., Edelson, S.G., Miller, R.L. *et al.* (1999) Contamination of intact apples after immersion in an aqueous environment containing *Escherichia coli* O157:H7. *Journal of Food Protection*, **62**(5), 444-450.

Buckner, P., Ferguson, D., Anzalone, F. *et al.* (1994) Outbreak of *Salmonella enteritidis* associated with a homemade ice cream – Florida, 1993. *Morbidity and Mortality Weekly Report,* **43**(36), 669–671.

Bunning, V.K., Crawford, R.G., Tierney, J.T. *et al.* (1990) Thermotolerance of *Listeria monocytogenes* and *Salmonella typhimurium* after sublethal heat shock. *Applied and Environmental Microbiology,* **56**(10), 3216–3219.

Burnett, S.L., Gehm, E.R., Weissinger, W.R. *et al.* (2000) A note: Survival of *Salmonella* in peanut butter and peanut butter spread. *Journal of Applied Microbiology,* **89**(3), 472–477.

Campden & Chorleywood Food Research Association (1997). *Manual of Microbiological Methods for the Food and Drinks Industry, Technical Manual No. 43.* Campden & Chorleywood Food Research Association, Chipping Campden, UK.

Carraminãna, J. J., Yangüela, J., Blanco, D. *et al.* (1997) *Salmonella* incidence and distribution of serotypes throughout processing in a Spanish poultry slaughterhouse. *Journal of Food Protection,* **60**(11), 1312–1317.

Chang, Y.H. (2000) Prevalence of *Salmonella* spp. in poultry broilers and shell eggs in Korea. *Journal of Food Protection,* **63**(5), 655–658.

Chilled Food Association (1997) *Guidelines for Good Hygienic Practice in the Manufacture of Chilled Foods,* 3rd edn.; (2001) High risk area – best practice guidelines, 2nd edn, Chilled Food Association, London.

Chung, K.C. and Goepfert, J.M. (1970) Growth of *Salmonella* at low pH. *Journal of Food Science,* **35**, 326–328.

C.I.M.S.C.E.E. (1992) *Code for the production of microbiologically safe and stable emulsified and non-emulsified sauces containing acetic acid.* Comité des Industries des Mayonnaises et Sauces Condimentaires de la Communauté Économique Européenne, Bruxelles, Belgium.

Codex Alimentarius Commission (1996a) Draft: *Revised Principles for the Establishment and Application of Microbiological Criteria for Foods,* Report of the twenty-ninth session of the Codex Committee on Food Hygiene, 21–25 October 1996, Washington, D.C. ALINORM 97/13A, Appendix III, Joint FAO/WHO Food Standards Programme. Codex Alimentarius Commission, Rome.

Codex Alimentarius Commission (1996b) *Establishment of sampling plans for microbiological safety criteria for foods in International trade including recommendations for the control of* Listeria monocytogenes, Salmonella enteritidis, Campylobacter *and entero-haemorrhagic* E. coli. Document prepared by the International Commission on Microbiological Specifications for Foods for the Codex Committee on Food Hygiene, twenty-ninth session, 21–25 October 1996, Agenda item 11, CX/FH 96/9 1–16. Codex Alimentarius Commission, Rome.

Cody, S.H., Abbott, S.L., Marfin, A.A. *et al.* (1999) Two outbreaks of multidrug-resistant *Salmonella* serotype Typhimurium DT104 infections linked to raw-milk cheese in Northern California. *Journal of the American Medical Association,* **281**(19), 1805–1810.

Cohen, D.R., Porter, I.A., Reid, T.M.S. *et al.* (1983) A cost benefit study of milk-borne salmonellosis. *Journal of Hygiene,* Cambridge, **91**, 17–23.

Collins-Thompson, D.L., Weiss, K.F., Riedel, G.W. *et al.* (1981) Survey of and microbiological guidelines for chocolate and chocolate products in Canada. *Canadian Institute of Food Science and Technology Journal,* **14**(3), 203–208.

Cook, K.A., Dobbs, T.E., Hlady, W.G. *et al.* (1998) Outbreak of *Salmonella* serotype Hartford infections associated with unpasteurized orange juice. *Journal of the American Medical Association,* **280**(17), 1504–1509.

Corrier, D.E., Nisbet, D.J., Byrd II, J.A. *et al.* (1998) Dosage titration of a characterized competitive exclusion culture to inhibit *Salmonella* colonization in broiler chickens during growout. *Journal of Food Protection*, **61**(7), 796–801.

Corry, J.E.L. (1974) The effect of sugars and polyols on the heat resistance of salmonellae. *Journal of Applied Bacteriology*, **37**, 31–43.

Corry, J.E.L., James, C., James, S.J. *et al.* (1995) *Salmonella, Campylobacter* and *Escherichia coli* O157:H7 decontamination techniques for the future. *International Journal of Food Microbiology*, **28**, 187–196.

Cowden, J.M., O'Mahony, M., Bartlett, C.L.R. *et al.* (1989) A national outbreak of *Salmonella typhimurium* DT124 caused by contaminated salami sticks. *Epidemiology and Infection*, **103**, 219–225.

Craven, P.C., Mackel, D.C., Baine, W.B. *et al.* (1975) International outbreak of *Salmonella eastbourne* infection traced to contaminated chocolate. *The Lancet*, 5 April 1975, 788–793.

Crespin, F.H., Eason, B., Gorbitz, K. *et al.* (1995) Outbreak of salmonellosis associated with beef jerky – New Mexico, 1995. *Morbidity and Mortality Weekly Report*, **44**(42), 785–788.

Cronin, M. (1999) Outbreak of *Salmonella enteritidis* associated with shell eggs in Ireland. *Eurosurveillance Weekly*, **46**(11 Nov), 2–3.

Dalsgaard, A. (1998) The occurrence of human pathogenic *Vibrio* spp. and *Salmonella* in aquaculture. *International Journal of Food Science and Technology*, **33**, 127–138.

Davies, R. (2000) *Salmonella* monitoring and control in pigs. Presented at *Zoonotic infections in livestock and the risk to human health*, London, UK. December 7, 2000.

D'Aoust, J.-Y. (1977) *Salmonella* and the chocolate industry. A review. *Journal of Food Protection,* **40**(10), 718–727.

D'Aoust, J.-Y. (1989) Salmonella *in Foodborne Bacterial Pathogens* (ed. M.P. Doyle). Marcel Dekker, Inc., New York, USA, pp. 327–445.

D'Aoust, J.-Y. (1991) Psychrotrophy and foodborne *Salmonella. International Journal of Food Microbiology*, **13**, 207–216.

D'Aoust, J.-Y. (1997) Salmonella *Species in Food Microbiology – Fundamentals and Frontiers* (eds M.P. Doyle, L.R. Beuchat and T.J. Montville). ASM Press, Washington, D.C., pp. 129–158.

D'Aoust, J.-Y., Aris, B.J., Thisdele, P. *et al.* (1975) *Salmonella eastbourne* outbreak associated with chocolate. *Journal of the Institute of Canadian Science and Technology Aliment*, **8**(4), 181–184.

D'Aoust, J.-Y., Emmons, D.B., McKellar, R. *et al.* (1987) Thermal inactivation of *Salmonella* species in fluid milk. *Journal of Food Protection*, **50**(6), 494–501.

D'Aoust, J.-Y., Warburton, D.W. and Sewell, A.M. (1985) *Salmonella typhimurium* phage type 10 from cheddar cheese implicated in a major Canadian foodborne outbreak. *Journal of Food Protection*, **48**(12), 1062–1066.

Delmore Jr, R.J., Sofos, J.N., Schmidt, G.R. *et al.* (2000) Interventions to reduce microbiological contamination of beef variety meats. *Journal of Food Protection*, **63**(1), 44–50.

de Louvois, J. (1993) *Salmonella* contamination of eggs: a potential source of human salmonellosis. *PHLS Microbiology Digest*, **10**(3), 158–162.

de Louvois, J. (1994) *Salmonella* contamination of stored hens' eggs. *PHLS Microbiology Digest*, **11**(4), 203–205.

de Louvois, J. and Rampling, A. (1998) One fifth of samples of unpasteurised milk are contaminated with bacteria. *British Medical Journal*, **316**, 625.

Desenclos, J-C., Bouvet, P., Benz-Lemoine, E. *et al.* (1996) Large outbreak of *Salmonella enterica* serotype *paratyphi* B infection caused by a goats' milk cheese, France, 1993: a case finding and epidemiological study. *British Medical Journal*, **312** (13 January), 91–94.

De Valk, H., Delarocque-Astagneau, E., Colomb, G. *et al.* (2000) A community-wide outbreak of *Salmonella enterica* serotype Typhimurium infection associated with eating a raw milk soft cheese in France. *Epidemiology and Infection*, **124**, 1–7.

Dodhia, H., Kearney, J. and Warburton, F. (1998) A birthday party, home-made ice cream, and an outbreak of *Salmonella enteritidis* phage type 6 infection. *Communicable Disease and Public Health*, **1**(1), 31–34.

Doyle, M.E. and Mazzotta, A.S. (2000) Review of studies on the thermal resistance of salmonellae. *Journal of Food Protection*, **63**(6), 779–795.

Ellajosyula, K.R., Doores, S., Mills, E.W. *et al.* (1998) Destruction of *Escherichia coli* O157:H7 and *Salmonella typhimurium* in Lebanon bologna by interaction of fermentation pH, heating temperature, and time. *Journal of Food Protection*, **61**(2), 152–157.

Ellis, A., Preston, M., Borczyk, A. *et al.* (1998) A community outbreak of *Salmonella berta* associated with a soft cheese product. *Epidemiology and Infection*, **120**, 29–35.

Ercolani, G.L. (1976) Bacteriological quality assessment of fresh marketed lettuce and fennel. *Applied and Environmental Microbiology*, **31**(6), 847–852.

European Council (1998) Council Directive 98/83/EC on the quality of water intended for human consumption. *Official Journal of the European Communities*, **L330**, 5.12.98, 32.

Evans, H.S., Madden, P., Douglas, C. *et al.* (1998) General outbreaks of infectious intestinal disease in England and Wales: 1995 and 1996. *Communicable Disease and Public Health*, **1**(3), 165–171.

Evans, M.R., Salmon, R.L., Nehaul, L. *et al.* (1999) An outbreak of *Salmonella typhimurium* DT170 associated with kebab meat and yoghurt relish. *Epidemiology and Infection*, **122**, 377–383.

Evans, S.J. (2000) *Salmonella* surveillance of cattle and sheep in GB. Presented at *Zoonotic infections in livestock and the risk to human health*, London, UK. 7 December 2000.

Ewing, W.H. (1972) The nomenclature of *Salmonella*, its usage and definitions for the three species. *Canadian Journal of Microbiology*, **18**, 1629–1637.

Fairbrother, R.W. (1938) *A text-book of medical bacteriology*. William Heinemann (Medical Books) Ltd, London, UK.

Fontaine, R.E., Cohen, M.L., Martin, W.T. *et al.* (1980) Epidemic salmonellosis from Cheddar cheese: surveillance and prevention. *American Journal of Epidemiology*, **111**(2), 247–253.

Food and Drug Administration (1998) Food labelling: Warning and notice statement: Labelling of juice products; Final rule. Hazard analysis and critical control point (HACCP); Procedures for the safe and sanitary processing and importing of juice; Extension of comment period; Proposed rule. *Code of Federal Regulations Parts 101 and 120, Department of Health and Human Services*, 8 July, 37029–37056. Food and Drug Administration, USA.

Food and Drug Administration (1999a) *Microbiological safety evaluations and recommendations on sprouted seeds*. National Advisory Committee on

Microbiological Criteria for Food. Adopted 28 May. Food and Drug Administration, USA.

Food and Drug Administration (1999b) Consumers advised of risks associated with raw sprouts. US Department of Health and Human Services. *HHS News*. 9 July. Food and Drug Administration, USA.

Food and Drug Administration (1999c) FDA issues nationwide health warning about Sun Orchard unpasteurized orange juice brand products. US Department of Health and Human Services. *HHS News*. 10 July. Food and Drug Administration, USA.

Food and Drug Administration (1999d) FDA issues health warning about Sun Orchard fresh squeezed unpasteurized orange juice. US Department of Health and Human Services. *HHS News*. 16 November. Food and Drug Administration, USA.

Food and Drug Administration (1999e) Hazard analysis and critical control point (HACCP); Procedures for the safe and sanitary processing and importing of juice; Availability of new data and information and reopening of comment period; Proposed rule. *Code of Federal Regulations Part 120, Department of Health and Human Services*, 23 November, 65669–65671. Food and Drug Administration, USA.

Food and Drug Administration (1999f) 1999 Food Code. Section 3.4, Subpart 3.401. Cooking. Food and Drug Administration, USA. www.cfsan.fda.gov

Food and Drug Administration (2001a) FDA survey of imported fresh produce. US Food and Drug Administration, Center for Food Safety and Applied Nutrition. www.cfsan.fda.gov

Food and Drug Administration (2001b) FDA publishes final rule to increase safety of fruit and vegetable juices. US Food and Drug Administration. *HHS News*. 18 January 2001. www.cfsan.fda.gov

Food and Drug Administration (2001c) Hazard analysis and critical control point (HACCP); Procedures for the safe and sanitary processing and importing of juice; Final rule. Code of Federal Regulations 21, Part 120, **66**(13), 6137–6202. www.cfsan.fda.gov

Gabis, D.A., Flowers, R.S., Evanson, D. *et al.* (1989) A survey of 18 dry dairy product processing plant environments for *Salmonella, Listeria* and *Yersinia*. *Journal of Food Protection*, **52**(2), 122–124.

Garcia-Villanova Ruiz, B., Galvez Vargas, R. and Garcia-Villanova, R. (1987) Contamination on fresh vegetables during cultivation and marketing. *International Journal of Food Microbiology*, **4**, 285–291.

George, S.M., Richardson, L.C.C., Pol, I.E. *et al.* (1998) Effect of oxygen concentration and redox potential on recovery of sublethally heat-damaged cells of *Escherichia coli* O157:H7, *Salmonella enteritidis* and *Listeria monocytogenes. Journal of Applied Microbiology*, **84**, 903–909.

Gibson, B. (1973) The effect of high sugar concentrations on the heat resistance of vegetative microorganisms. *Journal of Applied Bacteriology*, **36**, 365–376.

Gilbert, R.J., de Louvois, J., Donovan, T. *et al.* (2000) Guidelines for the microbiological quality of some ready-to-eat foods sampled at the point of sale. *Communicable Disease and Public Health*, **3**(3), 163–167.

Gill, O.N., Sockett, P.N., Bartlett, C.L.R. *et al.* (1983) Outbreak of *Salmonella napoli* infection caused by contaminated chocolate bars. *The Lancet*, 12 March, 574–577.

Goepfert, J.M. and Biggie, R.A. (1968) Heat resistance of *Salmonella typhimur-*

ium and *Salmonella senftenberg* 775W in milk chocolate. *Applied Microbiology*, **16**(12), 1939-1940.

Goepfert, J.M. and Chung, K.C. (1970) Behaviour of *Salmonella* during the manufacture and storage of a fermented sausage product. *Journal of Milk and Food Technology*, **33**, 185-191.

Goepfert, J.M., Iskander, I.K. and Amundson, C.H. (1970) Relation of the heat resistance of salmonellae to the water activity of the environment. *Applied Microbiology*, **19**(3), 429-433.

González-Hevia, M.A., Gutierrez, M.F. and Mendoza, M.C. (1996) Diagnosis by a combination of typing methods of a *Salmonella typhimurium* outbreak associated with cured ham. *Journal of Food Protection*, **59**(4), 426-428.

Goodfellow, S.J. and Brown, W.L. (1978) Fate of *Salmonella* inoculated into beef for cooking. *Journal of Food Protection,* **41**(8), 598-605.

Greenwood, M.H. and Hooper, W.L. (1983) Chocolate bars contaminated with *Salmonella napoli*: an infectivity study. *British Medical Journal*, **286**(6375), 1394.

Greenwood, M., Winnard, G. and Bagot, B. (1998) An outbreak of *Salmonella enteritidis* phage type 19 infection associated with cockles. *Communicable Disease and Public Health*, **1**(1), 35-37.

Groisman, E.A. and Ochman, H. (1997) How *Salmonella* became a pathogen. *Trends in Microbiology*, **5**(9), 343-349.

Groisman, E.A. and Ochman, H. (2000) The path to *Salmonella*. *ASM News*, **66**(1), 21-27.

Gustavsen, S. and Breen, O. (1984) Investigation of an outbreak of *Salmonella oranienburg* infections in Norway, caused by contaminated black pepper. *American Journal of Epidemiology*, **119**, 806-812.

Hammer, K.A., Carson, C.F. and Riley, T.V. (1999) Antimicrobial activity of essential oils and other plant extracts. *Journal of Applied Microbiology*, **86**, 985-990.

Harrison, J.A. and Harrison, M.A. (1996) Fate of *Escherichia coli* O157:H7, *Listeria monocytogenes* and *Salmonella typhimurium* during preparation and storage of beef jerky. *Journal of Food Protection,* **59**(12), 1336-1338.

Hedberg, C.W., Korlath, J.A., D'Aoust, J.-Y. *et al.* (1992) A multistate outbreak of *Salmonella javiana* and *Salmonella oranienburg* infections due to consumption of contaminated cheese. *Journal of the American Medical Association*, **268**(22), 3203-3207.

Heinitz, M.L., Ruble, R.D., Wagner, D.E. *et al.* (2000) Incidence of *Salmonella* in fish and seafood. *Journal of Food Protection*, **63**(5), 579-592.

Heinonen-Tanski, H., Niskanen, E.M., Salmela, P. *et al.* (1998) *Salmonella* in animal slurry can be destroyed by aeration at low temperatures. *Journal of Applied Microbiology*, **85**, 277-281.

Hellström, L. (1980) Food transmitted *S. enteritidis* epidemic in 28 schools. In: *Proceedings of the World Congress on Foodborne Infections and Intoxications.* Institute of Veterinary Medicine, West Berlin, pp. 397-400.

Hennessy, T.W., Hedberg, C.W., Slutsker, L. *et al.* (1996) A national outbreak of *Salmonella enteritidis* infections from ice cream. *The New England Journal of Medicine*, **334**(20), 1281-1286.

Himathongkham, S., Bahari, S., Riemann, H. *et al.* (1999a) Survival of *Escherichia coli* O157:H7 and *Salmonella typhimurium* in cow manure and cow manure slurry. *FEMS Microbiology Letters*, **178**, 251-257.

Himathongkham, S., Nuanualsuwan, S. and Riemann, H. (1999b) Survival of

Salmonella enteritidis and *Salmonella typhimurium* in chicken manure at different levels of water activity. *FEMS Microbiology Letters*, **172**, 159-163.

Hockin, J.C., D'Aoust, J.-Y., Bowering, D. *et al.* (1989) An international outbreak of *Salmonella nima* from imported chocolate. *Journal of Food Protection*, **52**(1), 51-54.

Hogue, A.T., Ebel, E.D., Thomas, L.A. *et al.* (1997) Surveys of *Salmonella enteritidis* in unpasteurized liquid egg and spent hens at slaughter. *Journal of Food Protection*, **60**(10), 1194-1200.

Horwitz, M.A., Pollard, R.A., Merson, M.H. *et al.* (1977) A large outbreak of foodborne salmonellosis on the Navajo Nation Indian Reservation, epidemiology and secondary transmission. *American Journal of Public Health*, **67**(11), 1071-1076.

Hou, H., Singh, R.K., Muriana, P.M. *et al.* (1996) Pasteurisation of intact shell eggs. *Food Microbiology*, **13**(2), 93-101.

Humphrey, T.J., Lanning, D.G. and Leeper, D. (1984) The influence of scald water pH on the death rates of *Salmonella typhimurium* and other bacteria attached to chicken skin. *Journal of Applied Bacteriology*, **57**, 355-359.

Humphrey, T.J., Whitehead, A., Gawler, A.H.L. *et al.* (1991) Numbers of *Salmonella enteritidis* in the contents of naturally contaminated hens' eggs. *Epidemiology and Infection*, **106**, 489-496.

ILSI (2000) Salmonella *Typhimurium definitive type (DT) 104: a multi-resistant* Salmonella. Report prepared under the responsibility of the International Life Sciences Institute, Europe: Emerging Pathogen Task Force. ILSI Europe, Brussels, Belgium.

Institute of Food Science and Technology (IFST) (1993) *Listing of Codes of Practice applicable to foods*. Institute of Food Science and Technology, London, UK.

Institute of Food Science and Technology (IFST) (1999) *Development and Use of Microbiological Criteria for Foods*. Institute of Food Science and Technology, London, UK.

International Commission on Microbiological Specifications for Foods (ICMSF) (1980) *Microbial Ecology of Foods, Volume 2: Food Commodities*. Academic Press, London, UK.

International Commission on Microbiological Specifications for Foods (ICMSF) (1986) *Microorganisms in Foods 2. Sampling for Microbiological Analysis: Principles and Specific Applications*, 2nd edn. University of Toronto Press, Canada.

International Commission on Microbiological Specifications for Foods (ICMSF) (1996) *Microorganisms in Foods 5. Microbiological Specifications of Food Pathogens*. Blackie Academic & Professional, London, UK.

International Commission on Microbiological Specifications for Foods (ICMSF) (1998) *Microorganisms in Foods 6. Microbial Ecology of Food Commodities*. Blackie Academic & Professional, London, UK.

Jaquette, C.B., Beuchat, L.R. and Mahon, B.E. (1996) Efficacy of chlorine and heat treatment in killing *Salmonella stanley* inoculated onto alfalfa seeds and growth and survival of the pathogen during sprouting and storage. *Applied and Environmental Microbiology*, **62**(6), 2212-2215.

Joce, R., O'Sullivan, D.G., Strong, C. *et al.* (1990) A national outbreak of *Salmonella gold-coast*. *Communicable Disease Report Weekly*, **4**, 3-4.

Jones, L. (2000) *Molecular methods in food analysis: principles and examples*. Key topics in Food Science and Technology, No. 1. Campden & Chorleywood Food Research Association, Chipping Campden, UK.

Joseph, C.A., Mitchell, E.M., Cowden, J.M. *et al.* (1991) A national outbreak of salmonellosis from yeast flavoured products. *Communicable Disease Report*, **1**, Review Number 2, R16–19.

Juneja, V.K. and Eblen, B.S. (2000) Heat inactivation of *Salmonella typhimurium* DT 104 in beef as affected by fat content. *Letters in Applied Microbiology*, **30**, 461–467.

Jung, Y.S. and Beuchat, L.R. (2000) Sensitivity of multidrug-resistant *Salmonella typhimurium* DT104 to organic acids and thermal inactivation in liquid egg products. *Food Microbiology*, **17**, 63–71.

Kapperud, G., Gustavsen, S., Hellesnes, I. *et al.* (1990) Outbreak of *Salmonella typhimurium* infection traced to contaminated chocolate and caused by a strain lacking the 60-Megadalton virulence plasmid. *Journal of Clinical Microbiology*, **28**(12), 2597–2601.

Killalea, D., Ward, L.R., Roberts, D. *et al.* (1996) International epidemiological and microbiological study of outbreak of *Salmonella agona* infection from a ready-to-eat savoury snack – England and Wales and the United States. *British Medical Journal*, **313**, 1105–1107.

Lecos, C. (1986) Of microbes and milk: probing America's worst *Salmonella* outbreak. *Dairy and Food Sanitation*, **6**(4), 136–140.

Lee, B.H., Kermasha, S. and Baker, B.E. (1989) Thermal, ultrasonic and ultraviolet inactivation of *Salmonella* in thin films of aqueous media and chocolate. *Food Microbiology*, **6**, 143–152.

Lehmacher, A., Bockemühl, J. and Aleksic, S. (1995) Nationwide outbreak of human salmonellosis in Germany due to contaminated paprika and paprika-powdered potato chips. *Epidemiology and Infection*, **115**, 501–511.

Le Minor, L. (1984) Genus 111. *Salmonella* Lignières 1900, 389[AL]. In: *Bergey's Manual of Systematic Bacteriology Volume 1* (eds N.R. Krieg and J.G. Holt). Williams & Wilkins, Baltimore, USA, pp. 427–458.

Le Minor, L. and Popoff, M.Y. (1987) Designation of *Salmonella enterica* sp. nov., nom. rev., as the type and only species of the genus *Salmonella*. *International Journal of Systematic Bacteriology*, **37**(4), 465–468.

Letellier, A., Messier, S. and Quessy, S. (1999) Prevalence of *Salmonella* spp. and *Yersinia enterocolitica* in finishing swine at Canadian abattoirs. *Journal of Food Protection*, **62**(1), 22–25.

Lewis, D.A., Paramathasan, R., White, D.G. *et al.* (1996) Marshmallows cause an outbreak of infection with *Salmonella enteritidis* phage type 4. *Communicable Disease Report Review*, **6**(13), R183–R186.

Licari, J.J. and Potter, N.N. (1970) *Salmonella* survival during spray drying and subsequent handling of skim milk powder. III Effects of storage temperature on *Salmonella* and dried milk properties. *Journal of Dairy Science*, **53**(7), 877–882.

Little, C.L., Monsey, H.A., Nichols, G.L. *et al.* (1997) The microbiological quality of cooked, ready-to-eat, out-of-shell molluscs. A report of the results of a study by the LACOTS/PHLS Co-ordinated Food Liaison Group, Microbiological Sampling Group. *PHLS Microbiology Digest*, **14**(4), 196–201.

Louie, K.K., Paccagnella, A.M., Osei, W.D. *et al.* (1993) *Salmonella* serotype Tennessee in powdered milk products and infant formula – Canada and United States, 1993. *Morbidity and Mortality Weekly Report*, **42**(26), 516–517.

Lundbeck, H., Plazikowski, U. and Silverstolpe, L. (1955) The Swedish *Salmonella* outbreak of 1953. *Journal of Applied Bacteriology*, **18**(3), 535–548.

Lyon, B.G., Berry, B.W., Soderberg, D. *et al.* (2000) Visual color and doneness indicators and the incidence of premature brown color in beef patties cooked to four end point temperatures. *Journal of Food Protection*, **63**(10), 1389–1398.

Mackey, B.M. and Derrick, C.M. (1986) Elevation of the heat resistance of *Salmonella typhimurium* by sublethal heat shock. *Journal of Applied Bacteriology*, **61**, 389–393.

Mackey, B.M. and Derrick, C.M. (1990) Heat shock protein synthesis and thermotolerance in *Salmonella typhimurium*. *Journal of Applied Bacteriology*, **69**, 373–383.

Maguire, H., Cowden, J., Jacob, M. *et al.* (1992) An outbreak of *Salmonella dublin* infection in England and Wales associated with a soft unpasteurised cows' milk cheese. *Epidemiology and Infection*, **109**, 389–396.

Mahon, B.E., Pönkä, A., Hall, W.N. *et al.* (1997) An international outbreak of *Salmonella* infections caused by alfalfa sprouts grown from contaminated seeds. *Journal of Infectious Diseases*, **175**, 876–882.

Mattick, K.L., Jørgensen, F., Legan, J.D. *et al.* (2000) Survival and filamentation of *Salmonella enterica* Serovar Enteritidis PT4 and *Salmonella enterica* Serovar Typhimurium DT104 at low water activity. *Applied and Environmental Microbiology*, **66**(4), 1274–1279.

McManus, C. and Lanier, J.M. (1987) *Salmonella, Campylobacter jejuni* and *Yersinia enterocolitica* in raw milk. *Journal of Food Protection*, **50**(1), 51–55.

Mead, G.C. (2000) Prospects for 'Competitive Exclusion' treatment to control salmonellas and other foodborne pathogens in poultry. *The Veterinary Journal*, **159**, 111–123.

Membré, J-M., Majchrzak, V. and Jolly, I. (1997) Effects of temperature, pH, glucose, and citric acid on the inactivation of *Salmonella typhimurium* in reduced calorie mayonnaise. *Journal of Food Protection*, **60**(12), 1497–1501.

Merker, R., Edelson-Mammel, S., Davis, V. *et al.* (1999) *Preliminary experiments on the effect of temperature differences on dye uptake by oranges and grapefruit*. US Food and Drug Administration. Center for Food Safety and Applied Nutrition, Washington. 23 November 1999.

Mitchell, E., O'Mahony, M., Lynch, D. *et al.* (1989) Large outbreak of food poisoning caused by *Salmonella typhimurium* definitive type 49 in mayonnaise. *British Medical Journal*, **298**, 14 January, 99–101.

Mokgatia, R.M., Brözel, V.S. and Gouws, P.A. (1998) Isolation of *Salmonella* resistant to hypochlorous acid from a poultry abattoir. *Letters in Applied Microbiology*, **27**, 379–382.

Morgan, D., Mawer, S.L. and Harman, P.L. (1994) The role of home-made ice cream as a vehicle of *Salmonella enteritidis* phage type 4 infection from fresh shell eggs. *Epidemiology and Infection*, **113**, 21–29.

Mossel, D.A.A., Corry, J.E.L., Struijk, C.B. and Baird, R.M. (1995) *Essentials of the Microbiology of Foods. A Textbook for Advanced Studies*. John Wiley & Sons Ltd, Chichester, UK.

Murphy, R.Y., Marks, B.P., Johnson, E.R. *et al.* (1999) Inactivation of *Salmonella* and *Listeria* in ground chicken breast meat during thermal processing. *Journal of Food Protection*, **62**(9), 980–985.

National Advisory Committee on Microbiological Criteria for Foods (1999) Microbiological safety evaluations and recommendations on sprouted seeds. *International Journal of Food Microbiology*, **52**, 123–153.

National Research Council (1985) *An evaluation of the Role of Microbiological Criteria for Foods and Food Ingredients*. US Subcommittee on

Microbiological Criteria, Committee on Food Protection, Food and Nutrition Board, National Research Council. National Academy Press, Washington, USA.

Natrajan, N. and Sheldon, B.W. (2000a) Efficacy of nisin-coated polymer films to inactivate *Salmonella* Typhimurium on fresh broiler skin. *Journal of Food Protection*, **63**(9), 1189–1196.

Natrajan, N. and Sheldon, B.W. (2000b) Inhibition of *Salmonella* on poultry skin using protein- and polysaccharide-based films containing nisin formulation. *Journal of Food Protection*, **63**(9), 1268–1272.

Ng, D.L.K., Koh, B.B., Tay, L. *et al.* (1999) The presence of *Salmonella* in local food and beverage items in Singapore. *Dairy, Food and Environmental Sanitation*, **19**(12), 848–852.

Ng, S., Rouch, G., Dedman, R. *et al.* (1996) Human salmonellosis and peanut butter. *Communicable Disease Intelligence*, **20**(14), 326.

Ng, W.F., Langlois, B.F. and Moody, W.G. (1997) Fate of selected pathogens in vacuum-packaged dry-cured (country-style) ham slices stored at 2°C and 25°C. *Journal of Food Protection*, **60**(12), 1541–1547.

Nicholson, F.A., Hutchison, M.L., Smith, K.A. *et al.* (2000) *A study of farm manure applications to agricultural land and an assessment of the risks of pathogen transfer into the food chain. Research Report.* Ministry of Agriculture, Fisheries and Food, UK.

Old, D.C. (1992) Nomenclature of *Salmonella*. *Journal of Medical Microbiology*, **37**, 361–363.

Old, D.C. and Threlfall, E.J. (1998) *Salmonella*. In: *Topley and Wilson's Microbiology and Microbial Infections*, 9th edn (eds A. Balows and B.I. Duerden). Arnold, London, UK, pp. 969–997.

O'Mahony, M., Cowden, J., Smyth, B. *et al.* (1990) An outbreak of *Salmonella saintpaul* infection associated with bean sprouts. *Epidemiology and Infection*, **104**, 229–235.

Orta-Ramirez, A., Price, J.F., Hsu, Y-C. *et al.* (1997) Thermal inactivation of *Escherichia coli* O157:H7, *Salmonella senftenberg*, and enzymes with potential as time–temperature indicators in ground beef. *Journal of Food Protection*, **60**(5), 471–475.

Palmer, S.R. and Rowe, B. (1986) Trends in *Salmonella* infections. *Public Health Laboratory Service Microbiology Digest*, **3**(2), 2–5.

Pao, S. and Davis, C.L. (1999) Enhancing microbiological safety of fresh orange juice by fruit immersion in hot water and chemical sanitizers. *Journal of Food Protection*, **62**(7), 756–760.

Parish, M.E. (1998) Coliforms, *Escherichia coli* and *Salmonella* serovars associated with a citrus-processing facility implicated in a salmonellosis outbreak. *Journal of Food Protection*, **61**(3), 280–284.

Parish, M.E., Narciso, J.A. and Friedrich, L.M. (1997) Survival of salmonellae in orange juice. *Journal of Food Safety*, **17**, 273–281.

Park, H.S., Marth, E.H., Goepfert, J.M. *et al.* (1970) The fate of *Salmonella typhimurium* in the manufacture and ripening of low-acid Cheddar cheese. *Journal of Milk and Food Technology*, **33**, 280–284.

Pennington, T.H. (1997) *The Pennington Group: Report on the Circumstances leading to the 1996 Outbreak of Infection with E. coli O157 in Central Scotland, the Implications for Food Safety and the Lessons to be Learned.* The Stationery Office Ltd, Edinburgh, Scotland.

Phillips, L.E., Humphrey, T.J. and Lappin-Scott, H.M. (1998) Chilling invokes different morphologies in two *Salmonella enteritidis* PT4 strains. *Journal of Applied Microbiology*, **84**, 820-826.

Plummer, R.A.S., Blissett, S.J. and Dodd, C.E.R. (1995) *Salmonella* contamination of retail chicken products sold in the UK. *Journal of Food Protection*, **58**(8), 843-846.

Pontello, M., Sodano, L., Nastasi, A. *et al.* (1998) A community-based outbreak of *Salmonella enterica* serotype Typhimurium associated with salami consumption in Northern Italy. *Epidemiology and Infection*, **120**, 209-214.

Puohiniemi, R., Heiskanen, T. and Siitonen, A. (1997) Molecular epidemiology of two international sprout-borne *Salmonella* outbreaks. *Journal of Clinical Microbiology*, **35**(10), 2487-2491.

Rajashekara, G., Haverly, E., Halvorson, D.A. *et al.* (2000) Multidrug-resistant *Salmonella* Typhimurium DT104 in poultry. *Journal of Food Protection*, **63**(2), 155-161.

Ray, B., Jezeski, J.J. and Busta, F.F. (1971a) Isolation of salmonellae from naturally contaminated dried milk products. II. Influence of storage time on the isolation of salmonellae. *Journal of Milk and Food Technology*, **34**(9), 423-427.

Ray, B., Jezeski, J.J. and Busta, F.F. (1971b) Isolation of salmonellae from naturally contaminated dried milk products. I. Influence of sampling procedure on the isolation of salmonellae. *Journal of Milk and Food Technology*, **34**(8), 389-393.

Rayman, M.K., D'Aoust, J.-Y., Aris, B. *et al.* (1979) Survival of microorganisms in stored pasta. *Journal of Food Protection*, **42**(4), 330-334.

Reporter, R., Mascola, L., Kilman, L. *et al.* (2000) Outbreaks of *Salmonella* serotype Enteritidis infection associated with eating raw or undercooked shell eggs – United States, 1996-1998. *Morbidity and Mortality Weekly Report*, **49**(4), 73-79.

Roberts, D. (1991) *Salmonella* in chilled and frozen chicken. *The Lancet*, 20 April, **337**, 984-985.

Roberts, D., Hooper, W. and Greenwood, M. (eds) (1995) *Practical Food Microbiology, Methods for the examination of food for micro-organisms of Public Health Significance*. Public Health Laboratory Service, London, UK.

Rowe, B. and Hall, M.L.M. (1989) *Kauffmann–White Scheme*. Division of Enteric Pathogens, Central Public Health Laboratory, London, UK.

Rowe, B., Hutchinson, D.N., Gilbert, R.J. *et al.* (1987) *Salmonella ealing* infections associated with consumption of infant dried milk. *The Lancet*, 17 October, 900-903.

Rushdy, A.A., Stuart, J.M., Ward, L.R. *et al.* (1998) National outbreak of *Salmonella senftenberg* associated with infant food. *Epidemiology and Infection*, **120**, 125-128.

Ryan, C.A., Nickels, M.K., Hargrett-Bean, N.T. *et al.* (1987) Massive outbreak of antimicrobial-resistant salmonellosis traced to pasteurised milk. *Journal of the American Medical Association*, **258**(22), 3269-3274.

Sadik, Ch., Krending, M.-J., Mean, F. *et al.* (1986) An epidemiological investigation following an infection by *Salmonella typhimurium* due to the ingestion of cheese made from raw milk. In: *Proceedings of the 2nd World Congress on Foodborne Infections and Intoxications, 1*, Berlin, pp. 280-282.

Sauer, C.J., Majkowski, J., Green, S. *et al.* (1997) Foodborne illness outbreak associated with a semi-dry fermented sausage product. *Journal of Food Protection*, **60**(12), 1612-1617.

Savage, W.G. (1912) *Milk and the Public Health*. Macmillan and Co. Ltd, London, UK.

Schroeder, S.A. (1968) What the sanitarian should know about salmonellae and staphylococci in milk and milk products. *Journal of Milk and Food Technology*, **30**, 376-380.

Shapiro, R., Ackers, M-L., Lance, S. *et al.* (1999) *Salmonella* Thompson associated with improper handling of roast beef at a restaurant in Sioux Falls, South Dakota. *Journal of Food Protection*, **62**(2), 118-122.

Shohat, T., Green, M.S., Merom, D. *et al.* (1996) International epidemiological and microbiological study of outbreak of *Salmonella agona* infection from a ready-to-eat savoury snack – II: Israel. *British Medical Journal*, **313**, 1107-1109.

Sin, J., Quigley, C. and Davies, M. (2000) Survey of raw egg use by home caterers. *Communicable Disease and Public Health*, **3**(2), 90-94.

Skov, M.N., Carstensen, B., Tornøe, N. *et al.* (1999) Evaluation of sampling methods for the detection of *Salmonella* in broiler flocks. *Journal of Applied Microbiology*, **86**, 695-700.

Slavik, M.F., Kim, J-W. and Walker, J.T. (1995) Reduction of *Salmonella* and *Campylobacter* on chicken carcasses by changing scalding temperature. *Journal of Food Protection*, **58**(6), 689-691.

Smith, J.L., Huhtanen, C.N., Kissinger, J.C. *et al.* (1975) Survival of salmonellae during pepperoni manufacture. *Applied Microbiology*, **30**(5), 759-763.

Sockett, P.N., Cowden, J.M., Le Baigue, S. *et al.* (1993) Foodborne disease surveillance in England and Wales: 1989-1991. *Communicable Disease Report Review*, **3**(12), R164.

Sofos, J.N., Kochevar, S.L., Reagan, J.O. *et al.* (1999) Incidence of *Salmonella* on beef carcasses relating to the U.S. meat and poultry inspection regulations. *Journal of Food Protection*, **62**(5), 467-473.

Synnott, M.B., Brindley, M., Gray, J. *et al.* (1998) An outbreak of *Salmonella agona* infection associated with precooked turkey meat. *Communicable Disease and Public Health*, **1**(3), 176-179.

Tamblyn, K.C. and Conner, D.E. (1997) Bactericidal activity of organic acids in combination with transdermal compounds against *Salmonella typhimurium* attached to broiler skin. *Food Microbiology*, **14**, 477-484.

Tamminga, S.K. (1979) The longevity of *Salmonella* in chocolate. *Antonie van Leeuwenhoek*, **45**, 153-156.

Tamminga, S.K., Beumer, R.R., Kampelmacher, E.H. *et al.* (1976) Survival of *Salmonella eastbourne* and *Salmonella typhimurium* in chocolate. *Journal of Hygiene*, Cambridge, **76**, 41-47.

Taormina, P.J., Beuchat, L.R. and Slutsker, L. (1999) Infections associated with eating seed sprouts: An international concern. *Emerging Infectious Diseases*, **5**(5), 626-640.

Taplin, J. (1982) *Salmonella newport* outbreak – Victoria. *Communicable Disease Intelligence*, **1**, 3-5.

Tauxe, R., Kruse, H., Hedberg, C. *et al.* (1997) Microbial Hazards and Emerging Issues Associated with Produce. A Preliminary Report to the National Advisory Committee on Microbiologic Criteria for Foods. *Journal of Food Protection*, **60**(11), 1400-1408.

Taylor, J.P., Shandera, W.X., Betz, T.G. *et al.* (1984) Typhoid fever in San Antonio, Texas: An outbreak traced to a continuing source. *Journal of Infectious Diseases*, **149**(4), 553-557.

Thompson, S.S., Harmon, L.G. and Stine, C.M. (1978) Survival of selected organisms during spray drying of skim milk and storage of non-fat dry milk. *Journal of Food Protection*, **41**(1), 16-19.

Threlfall, E.J., Hall, M.L.M. and Rowe, B. (1983) Lactose-fermenting salmonellae in Britain. *FEMS Microbiology Letters*, **17**, 127-130.

Threlfall, J., Ward, L. and Old, D. (1999) Changing the nomenclature of *Salmonella*. *Communicable Disease and Public Health*, **2**(3), 156-157.

Topley, W.W.C. and Wilson, G.S. (1929a) *The Principles of Bacteriology and Immunity*, Volume I. Edward Arnold and Co., London, UK, pp. 429 and 445.

Topley, W.W.C. and Wilson, G.S. (1929b) *The Principles of Bacteriology and Immunity*, Volume II. Edward Arnold and Co., London, UK, pp. 1037-1038.

Trepka, M.J., Archer, J.R., Altekruse, S.F. *et al.* (1999) An increase in sporadic and outbreak-associated *Salmonella* Enteritidis infections in Wisconsin: The role of eggs. *Journal of Infectious Diseases*, **180**, 1214-1219.

Uljas, H.E. and Ingham, S.C. (1999) Combinations of intervention treatments resulting in 5-log$_{10}$ unit reductions in numbers of *Escherichia coli* O157 : H7 and *Salmonella typhimurium* DT104 organisms in apple cider. *Applied and Environmental Microbiology*, **65**(5), 1924-1929.

Uyttendaele, M., De Troy, P. and Debevere, J. (1999) Incidence of *Salmonella*, *Campylobacter jejuni*, *Campylobacter coli* and *Listeria monocytogenes* in poultry carcasses and different types of poultry products for sale on the Belgian retail market. *Journal of Food Protection*, **62**(7), 735-740.

Uzzau, S., Brown, D.J., Wallis, T. *et al.* (2000) Review: Host adapted serotypes of *Salmonella enterica*. *Epidemiology and Infection*, **125**, 229-255.

Vaillant, V., Haeghebaert, S., Desenclos, J.C. *et al.* (1996) Outbreak of *Salmonella dublin* infection in France, November-December 1995. *Eurosurveillance*, **1**(2), 9-10.

Van Beneden, C.A., Keene, W.E., Strang, R.A. *et al.* (1999) Multinational outbreak of *Salmonella enterica* serotype Newport infections due to contaminated alfalfa sprouts. *Journal of the American Medical Association*, **281**(2), 158-162.

Veeramuthu, G.J., Price, J.F., Davis, C.E. *et al.* (1998) Thermal inactivation of *Escherichia coli* O157 : H7, *Salmonella senftenberg*, and enzymes with potential as time-temperature indicators in ground turkey thigh meat. *Journal of Food Protection*, **61**(2), 171-175.

Vought, K.J. and Tatini, S.R. (1998) *Salmonella enteritidis* contamination of ice cream associated with a 1994 multi-state outbreak. *Journal of Food Protection*, **61**(1), 5-10.

Walker, W. (1965) The Aberdeen typhoid outbreak of 1964. *Scottish Medical Journal*, **10**, 466-479.

Wang, W-C., Li, Y., Slavik, M.F. *et al.* (1997) Trisodium phosphate and cetylpyridinium chloride spraying on chicken skin to reduce attached *Salmonella typhimurium*. *Journal of Food Protection*, **60**(8), 992-994.

Weissinger, W.R. and Beuchat, L.R. (2000) Comparison of aqueous chemical treatments to eliminate *Salmonella* on alfalfa sprouts. *Journal of Food Protection*, **63**(11), 1475-1482.

Wells, S.J., Fedorka-Cray, P.J., Dargatz, D.A. *et al.* (2001) Fecal shedding of *Salmonella* spp. by dairy cows on farm and at cull cow markets. *Journal of Food Protection*, **64**(1), 3-11.

Wheeler, J.G., Cowden, J.M., Sethi, D. *et al.* (1999) Study of infectious intestinal disease in England; rates in the community, presenting to GPs, and reported to national surveillance. *British Medical Journal*, **318**, 1046-1050.

Wierup, M., Engström, B., Engvall, A. *et al.* (1995) Control of *Salmonella enteritidis* in Sweden. *International Journal of Food Microbiology*, **25**, 219-226.

Wilson, G.S. (1942) *The pasteurization of milk*. Edward Arnold and Co., London, UK.

Wilson, G.S. and Miles, A.A. (1964) *Topley and Wilson's Principles of Bacteriology and Immunity*, Volume 1, 5th edn. Edward Arnold (Publishers), London, pp. 866–911, and Volume II, 1833–1866.

Wilson, I.G., Heaney, J.C.N. and Powell, G.G. (1998) *Salmonella* in raw shell eggs in Northern Ireland: 1996–7. *Communicable Disease and Public Health*, **1**(3), 156–160.

World Health Organisation (1988) *Salmonellosis Control: the role of animal and product hygiene*. Report of a WHO Expert Committee. Technical Report Series 774. World Health Organisation, Geneva, Switzerland.

Wright, J.R., Sumner, S.S., Hackney, C.R. *et al.* (2000a) A survey of Virginia apple cider producers' practices. *Dairy, Food and Environmental Sanitation*, **20**(3), 190–195.

Wright, J.R., Sumner, S.S., Hackney, C.R. *et al.* (2000b) Reduction of *Escherichia coli* O157 : H7 on apples using wash and chemical sanitizer treatments. *Dairy, Food and Environmental Sanitation*, **20**(2), 120–126.

Xavier, I.J. and Ingham, S.C. (1997) Increased *D*-values for *Salmonella enteritidis* following heat shock. *Journal of Food Protection*, **60**(2), 181–184.

Yang, Z., Li, Y. and Slavik, M. (1998) Use of antibacterial spray applied with an inside-outside birdwasher to reduce bacterial contamination on prechilled chicken carcasses. *Journal of Food Protection*, **61**(7), 829–832.

Zhuang, R-Y., Beuchat, L.R. and Angulo, F.J. (1995) Fate of *Salmonella montevideo* on and in raw tomatoes as affected by temperature and treatment with chlorine. *Applied and Environmental Microbiology*, **61**(6), 2127–2131.

INDEX